T0309472

Nature's Longest Threads

New Frontiers in the
Mathematics and Physics
of Information in Biology

Nature's Longest Threads

Threads

New Frontiers in the
Mathematics and Physics
of Information in Biology

Editors

Janaki Balakrishnan
B V Sreekantan

Indian Institute of Science, India

 **World Scientific**

NEW JERSEY · LONDON · SINGAPORE · BEIJING · SHANGHAI · HONG KONG · TAIPEI · CHENNAI

Published by

World Scientific Publishing Co. Pte. Ltd.

5 Toh Tuck Link, Singapore 596224

USA office: 27 Warren Street, Suite 401-402, Hackensack, NJ 07601

UK office: 57 Shelton Street, Covent Garden, London WC2H 9HE

Library of Congress Cataloging-in-Publication Data
Nature's longest threads : new frontiers in the mathematics and physics of information in biology /
editors Janaki Balakrishnan (National Institute of Advanced Studies, India), B.V. Sreekantan
(National Institute of Advanced Studies, India).
    pages cm
  Includes bibliographical references and index.
  ISBN 978-9814612463 (hardcover : alk. paper)
    1. Quantum theory--Mathematics. 2. Biological systems--Measurement. 3. Information theory in
biology. I. Balakrishnan, Janaki, editor of compilation. II. Sreekantan, B. V. (Badanaval Venkata),
1925–   editor of compilation.
  QC174.17.M35N38 2014
  570.1'154--dc23

                                                                    2014014297

British Library Cataloguing-in-Publication Data
A catalogue record for this book is available from the British Library.

Printed in Singapore

Preface

Humankind has taken tremendous strides in furthering the frontiers of science and technology. We have sent spaceprobes into unexplored regions of the cosmos, sequenced the human genome to a high degree of accuracy, devised printers that generate 3D images, proposed the existence of a single force unifying the four fundamental forces in nature and have even experimentally confirmed the veracity of the standard electroweak hypothesis through the discoveries of the W and Z bosons and then more recently, in the signatures of the Higgs particle. Yet, we have so far not made adequate progress in understanding how our mind and cognitive faculties work.

Receiving inputs from the environment, how do thoughts and feelings form? How does one quantify consciousness and how is it linked to the physical laws and the material stuff out of which we are made? A quantum mechanical description would seem a natural choice since quantum laws rule the heart of matter, at the atomic scale and within. One could therefore describe information and its transfer through a probabilistic description on a Hilbert space or on a Bloch sphere, information being gleaned through quantum measurements. Ruled by the principle of superposition of states, the quantum world however, presents difficulties in the interpretation of measurements of a property in a certain state.

In living systems, information is conveyed through the network of sensory neurons and motor neurons/central nervous system to the brain. Patterns of electrical trains and bursts encode this information, so attempts at understanding the generation of these electrical bursts and patterns by the activity of coupled neurons, represent definitive steps forward in the study of information transfer at the cellular scale. The presynaptic neuron's chemical synapse, with its distribution of excitatory or inhibitory neurotransmitters packed in vesicles, all waiting to be released at the synaptic cleft, is a highly regulated, beautifully organised dynamical system. Why has Nature retained the electrical synapse? Is it merely for achieving faster synchrony in neuronal firing? The equations seeking to mimic the dynamic behaviour of neurons and their activity, describe the physical system at the classical level — a very nonlinear regime.

How then does one achieve the interface linking the classical information described by electrical signals with the conscious states of information of the individual, possibly describable by quantum information? When one makes a "conscious decision" about something, do the cognitive faculties then convert the probabilistic description entirely into a deterministic one? While on the one hand, challenging questions like these remain to be answered, on the other, it is known that quantum effects do manifest in living systems. An example is the detection of odors by the olfactory organ.

Physical models depend upon mathematical techniques to test their predictions. One could then ask if mathematics itself is entirely a fabrication of the human mind, or whether it exists independent of humankind. Some chapters in this volume have given very interesting and diverse views on this subject.

Nature's diversity and intricate beauty at each scale — from dimensions beyond the Planck scale to cosmic lengths encompassing the celestial orb — only teaches us that information and its mode of transfer may be of varying kinds across different regimes. In a living being likewise, thought processes linking the cognitive faculties with the "conscious" self represent modes of information transfer not yet understood by physical laws. As Richard Feynman observed in "The Character of Physical Law":

> "*Nature uses only the longest threads to weave her patterns, so that each small piece of her fabric reveals the organization of the entire tapestry*".

This volume is thus a compendium of differing views on different levels of information in living systems describable by physics and mathematics.

Several of these articles are contributions by speakers at a Workshop on "Application of Mathematics and Physics to Cognition and Consciousness" held 11–12 March, 2013 at the National Institute of Advanced Studies, Indian Institute of Science Campus, Bangalore. We hope the readers would find the material in the chapters as interesting and enjoyable as the Editors did.

Janaki Balakrishnan   Bangalore, India
B.V. Sreekantan   February 2014

Acknowledgements

We would like to acknowledge enthusiastic support from the World Scientific Publishing Company Commissioning Editor, Ms. Ranjana Rajan. She has been very helpful and made our interactions with the publisher a very pleasant experience. We would also like to thank Mr. Alvin Chong, Desk Editor, and other members of the World Scientific Publishing Team who are involved in the production of this edited volume, in particular, Mr. Rajesh Babu. We are grateful to Mrs. V.B. Mariyammal for help with LaTeX in the earlier stages of this edited volume. Finally we would like to acknowledge encouragement from Dr. V.S. Ramamurthy, Director, National Institute of Advanced Studies, Bangalore for organizing the Workshop which initiated the effort to compile this volume.

Contents

Chapter 1

Mathematics In-forms physics and physics Per-forms mathematics: comments

N. Kumar

Raman Research Institute C V Raman Avenue,
Bangalore 560080, India
nkumar@rri.res.in

In this conversation I hold that Mathematics Informs Physics and Physics Performs Mathematics accordingly. The two appear to be indifferent to and ignorant of the Human Condition (consciousness/self-awareness). And this Ignorance seems Invincible. I hope to illustrate some of these with a simple example taken from the mathematics of finite part of infinity (or, more precisely a regularization) based on Ramanujan, and the physics of the experimentally measured dispersion force (between two plane-parallel metallic plates) based on the experiment of Casimir: This measured Casimir force turns out to be in conformity with that calculated from Ramanujan's highly counter-intuitive finite part of infinity. Behold! the Unreasonable Effectiveness of Mathematics in Natural Sciences that Wigner had spoken of. But clearly, the first-person singular CONSCIOUSNESS is missing from this discourse which lives in the public domain.

Mathematics and Physics of the Casimir Effect (a simplified version)

- Consider the sum of all natural numbers:
  $S = 1 + 2 + 3 + 4 + \cdots$ up to infinity (∞).
  (essentially, this is the infinite sum that occurs as an overall dimensionful multiplicative factor in the calculation of the Casimir force that has now been measured to high degree of accuracy experimentally).
- QUESTION:
  What is the numerical value of this sum?

- ANSWER:
    (a) the sum is $= \infty$ (obviously, says commonsense).
    (b) the sum is $= -1/12$ (admissible mathematically,
                                    would say Ramanujan)
    (c) the sum is $\cong -1/12$ (determined experimentally,
                                    would say Casimir).
- Turns out that physics has actually chosen one of the possible worlds (c)
    allowed by mathematics (b), which is experimentally indeed the
    correct one — though highly counter-intuitive.
    And this seems to occur all too often in the physical world of our
    public discourse that we inhabit.
- Below, I will try to give you a more explicit demonstration of this rela-
    tionship between physics and mathematics.
- The Casimir Experiment: Force between two plane-parallel metallic
    plates at distance d apart in vacuum. See Fig. 1.
- Commonsense says: Add up the energies of all such modes lying in the
    space between the two plates. This is a trivial exercise: The Casimir
    force is then readily shown to be proportional to just this sum of
    natural integers :

$$S = 1 + 2 + 3 + 4 + \cdots \text{up to infinity}(\infty),$$

WHICH IS CLEARLY INFINITE! NOT ACCEPTABLE!
(NOTE: The constant of proportionality in the force expression is
a known function of the plate spacing d, and involves fundamental
constants $\hbar, e$, and c. Here we have dropped this proportionality
constant for simplicity).

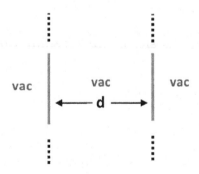

Fig. 1. Casimir force $F = -\frac{d}{dx}(E(x))$ at $x = d$, where $E(x)$ is the total vacuum
energy of the system (the so-called zero-point energy), which is the energy of all the
electromagnetic modes for the system in the ground state.

- What has then gone wrong?

  Our commonsense has just missed out an important contribution to the total system energy, which is also infinite but is opposite in sign: this comes from the realization that there is space also on the outside of the two plates, and it extends up to infinity. This too contains infinitely many modes with total energy $= \infty$. Thus, as the plate separation is changed ($\equiv \frac{d}{dx}$), the internal contribution $E(x)$ as well as the external contribution $E(x > d)$ both change: both are infinite but have the opposite signs. It is this cancellation of the two infinities that gives ultimately the finite answer for the net Casimir force which has the value as experimentally measured now.

- NOW, BEHOLD THE UNREASONABLE EFFECTIVENESS OF MATHEMATICS IN THE NATURAL SCIENCES!

  Turns out that it is sufficient to consider just the mode energies for the interior of the two-plate system only (ignoring completely the outlying infinite space). As pointed out earlier, this sum S is, of course, infinite $= 1 + 2 + 3 + \cdots + \cdots$ up to ∞.

  But, mathematics (Ramanujan) allows this infinite sum to have a finite part, which is just right

$$= -\frac{1}{12}$$

  as actually confirmed by the laboratory experiments (Casimir).

- Ramanujan's argument would be along the following lines:

  Define a convergent series

$$S_k = \sum_0^\infty n \times e^{-nk} \equiv -\sum_0^\infty \frac{\partial}{\partial k} e^{-kn} \quad \text{with k positive.}$$

  Note that our original sum $S = S_k$ in the limit the auxiliary parameter k tends to zero. This sum (S_k) is clearly finite, and can be trivially evaluated. All we have to do now is to expand this sum as a power series in the parameter k:

$$S_k = (\cdots)\frac{1}{k^4} + (\cdots)\frac{1}{k^3} + \left(-\frac{1}{12}\right)k^0 + \cdots \text{higher power of } k,$$

  and pick up the term k^0 which is independent of the auxiliary parameter k. This then gives us the *finite part of infinity* to be $-1/12$!

With this, we finally have the *result*,

$$\frac{\text{Casimir force}}{\text{plate area}} = -\frac{1}{12}\left(\frac{\pi^2}{20}\frac{\hbar c}{x^4}\right),$$

which is essentially correct. (Here, we have re-introduced the dimensionful coefficient of proportionality dropped earlier).

Now, how come the mathematics of Ramanujan *knows* about the physics of the Casimir effect, which involves the fact that the space (the Universe) outside of the two plates is infinite, and should also count towards the reckoning of the force of Casimir. **There seems to be a deep understanding here between physics and mathematics? That is the question.**

1. Appendix

Some details of this calculation of the Dispersion Force of Casimir between Two Perfectly Conducting Plane-Parallel Plates are calculated from textbook method. See Fig. 2.

Note that the *empty* space is really filled with fluctuating electromagnetic fields. The zero-point energy is given by

$$\Delta E(d) = E(d) + E(L-d) - E_L, \quad \text{where}$$

$$E(d) = 2\sum_{n=0}^{\infty}\theta_n\left(\frac{L^2}{\pi^2}\right)\int\int_0^\infty \tfrac{1}{2}\hbar\omega_{nk}\,d^2k$$

$$E(L-d) = \frac{2L^2(L-a)}{\pi^3}\int\int\int_0^\infty \tfrac{1}{2}\hbar ck\,d^3k$$

$$E_L = \left(\frac{2L^3}{\pi^3}\right)\int\int\int_0^\infty \tfrac{1}{2}\hbar ck\,d^3k,$$

where, $\theta_n = 1/2$ for $n = 0$ and $\theta_n = 1$ for $n > 0$. Note the three parts of $\Delta E(d)$ involved here. See Ref. 1.

With this,

$$\frac{\text{Force}}{\text{area}} \equiv -\frac{1}{L^2}\frac{\partial}{\partial d}E(d) = -\frac{\pi^2}{240}\left(\frac{\hbar c}{d^4}\right).$$

For example, for $A = 1cm^2$, $d = 0.5\mu m$, this is of the order of force of attraction between e^- and p^+ in H-atom, also of the order of gravitational attraction between 1 Kg masses 1 cm apart. See Ref. 2.

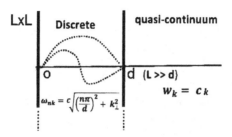

Fig. 2. Dispersion force of Casimir between two perfectly conducting plane-parallel plates.

Consciousness

- Does the will-free Public Domain of Science Exhaust the Universe of our Discourse?

  OR

  Is there a Private Domain of free-willed Consciousness too — the primary experience of self-awareness?

  Do these two domains overlap (interact)?

  THESE ARE THE QUESTIONS WE MUST ASK, AND POSSIBLY ANSWER.

- By science here, of course, I mean all of physical sciences — physics, mathematics, chemistry, biology,...; and the force laws that mediate their interactions. It is will-free. It can be shared. It includes the BRAIN too with all its

  Hardware $\sim 10^{11}$ neurons

                 $\sim 10^{14}$ synapses

                 ~ 20 Watt ... ,

  Software $\sim$ neural network

                 $\sim$ computation ...

- Consciousness in its private domain is free-willed, inalienable, self-referential. It cannot communicate, in and of itself, with the public domain of sciences.

  But, however, it also has the property of being *'self aware'*!

- Consciousness cannot be shared — indeed, we know from the split-brain experiments (involving dissection of the corpus callosum) that the thus separated hemispheres of the same individual suggest two different streams of consciousness.

- It could well be that the Public Domain of Physics and the Private Domain of Consciousness are governed by fundamentally different laws, *e.g.*, both may obey quantum mechanics, but with different Planck constants, $\hbar \neq \hbar'$, say. This, however, is disallowed — it leads to absurdity, for example, of continual dissipation.

- Attempts have been made to introduce interactions between the two domains through biocentric interpretation of some recent baffling experiments in physics, e.g. the delayed-choice experiments. I suspect these interpretations.

- Some interactions between the two domains are, however, expected to exist on general grounds: After all, in the beginning, following the Big Bang, there was nothing but inanimate entities — we certainly were not there! Evolution eventually led to a transition to the animate world of today that we inhabit. It is only natural then that this commonality of shared evolution should have its imprint marked on us — thus providing for a unifying degree of interaction (communication) among these two climax domains — the private and the public.

- It is, of course, not unthinkable, that the domain of science and that of consciousness may have between them no detectable interaction or communication whatsoever. If that be the case, then **this will be an IGNORANCE of Science and the IGNORANCE will be INVINCIBLE!**

References

1. Ballentine, L.E., *Quantum Mechanics: A Modern Development*. World Scientific, Singapore (1998) p.536.
2. Lamoraux, S.K.: 1997 *Demonstration of the Casimir Force in the* 0.6 *to* 6μm *Range*. Phys. Rev. Lett. **78**: 5-8.
3. Pippard, A.B.: 1988 *The Invincible Ignorance of Science*. Contemporary Physics **29**: 393-405.
4. Feynman, R.P. et al.: *Feynman lectures*. Reading: Massachusetts, Addison-Wesley, 1963, p.2.
5. Crick, F.: 1995 *The astonishing hypothesis*. London: Touchstone Books.
6. Wigner, E.P.: 1961 *The unreasonable effectiveness of mathematics in the natural sciences*. The Scientist Speculates, ed. I.J. Good, London: Heinemann.
7. Campbell, L. and Garnett, W.: 1882, *The life of James Clerk Maxwell*. Macmillan, London. Reprinted by Johnson Reprint, New York (1969) p. 440.
8. Lajos Diósi: *Quantum Dynamics with Two Planck Constants and the Semi-Classical Limit*. arXiv:quant-ph/9503023v1, 28 Mar 1995.
9. Kumar, N.: 2004 *Free-willed consciousness in interaction with will-free cosmos: Is there an inconsistency?* (Science and Beyond), eds. S. Menon, B.V. Sreekantan, A Sinha, P Clayton, and R Narasimha. (NIAS, Bangalore).

Chapter 2

An incomplete summing up of Quantum Measurements

N.D. Hari Dass

Chennai Mathematical Institute, Kelambakkam, Chennai 603103
CQIQC, Indian Institute of Science, Bangalore 560012
ndhari.dass@gmail.com

An attempt is made to give a flavor of the many important aspects of *Quantum Measurements*. I review here the genesis of the so called rules of quantum measurements and the Dirac–von Neumann model to set the stage for a discussion of a variety of quantum measurements that have appeared since then. Of these, I have discussed: (i) the Arthurs–Kelly simultaneous measurement of canonically conjugate pairs, (ii) the Protective measurements, (iii) the weak measurements, and finally, (iv) the weak value measurements.

1. Introduction

Measurements bridge theories with physical reality. Even a century after quantum theory, quantum measurements remain poorly understood. Discovery of quantum phenomena in other areas like biology may aid in their better understanding. Likewise, a better understanding of quantum measurements may accelerate identifying such phenomena too.

A classical system is described by its *state*. There are *properties* of the system, and their *values* depend on the state of the system. Properties of a system in any given state can be *measured*. The measurement process gives well defined values, usually in some range, determined by the *resolution*. The classical state is *not disturbed* by the measurement. *All* properties of the system can be *simultaneously* determined. With time, the state changes *deterministically*, according to some well defined laws, for example, Newton's laws. *No additional rules are required for describing measurements.*

Quantum systems are also described by their *states*. They too can possess many *properties*. As an example, consider the quantum analogue of the bits of classical computing, called *qubits*. The list of properties in this case are called $\sigma_x, \sigma_y, \sigma_z$ and any linear combination of them with *real* coefficients. In quantum theory properties are *matrices* (more generally, *operators*). In the qubit case they are 2×2 matrices with *eigenvalues* $+1$ and -1. As per the standard lore, only *eigenstates* of a property, say, σ_z, can have *values* of the said property, and the value is the corresponding *eigenvalue*.

Let us first focus on the property σ_z (it is of course immaterial as to which of the properties one chooses at this stage). Call $|+1\rangle$ the eigenvector with eigenvalue $+1$, and likewise for the one with eigenvalue -1. At this stage the situation is the same as in the classical bit case. The two eigenvectors are of unit length, and orthogonal to each other.

2. The Principle of Superposition — A Bolt from the Blue

Now comes one of the biggest shocks from quantum theory called *the Principle of Superposition of States*. As stressed by Dirac,[1] *it is like no other superposition principle in physics!* It says, if $|\psi_1\rangle$ is a quantum state, and $|\psi_2\rangle$ another state, then *infinitely many other states* can be constructed from these through *superposition*.

When applied to the eigenstates of σ_z, this principle, for example, asserts that $|\chi\rangle = \alpha\,|+1\rangle + \beta\,|-1\rangle$ are *all* bona fide states when $|\alpha|^2 + |\beta|^2 = 1$.

What precisely is the meaning of such a superposition? In the general context of scattering, this was seen as essentially similar to the superposition principle in optics. But as can be seen from the qubit example, superposition can be between states with *classically conflicting* values of a property. The true implications of this principle were dramatized on the one hand by Schrödinger through his famous *Schrödinger Cat Paradox*, and on the other hand by Einstein through the famous *EPR Paradox*.

The essence of the difficulties can be brought out by the precise question: *"What will be the outcome of a measurement of σ_z in the state $|\chi\rangle$ (for given α, β)?"* Quantum theory was rescued from this deep dilemma by the *Probability Interpretation* of Born. The joint wisdom of Dirac, Born and von Neumann[1,2] led to the following *rules* of quantum measurements: If we perform the measurement of σ_z on a *single copy* of $|\chi\rangle$, the outcome is one of the eigenvalues, i.e. $+1$ or -1, *completely at random*. If the outcome is $+1$, the state of the system after measurement is $|\sigma_z = 1\rangle$, and if it is -1 the post-measurement state is $|\sigma_z = -1\rangle$. If measurements are performed on a large ensemble of N identical copies of $|\chi\rangle$, the outcome $+1$ occurs with probability $|\alpha|^2$ and outcome -1 with probability $|\beta\rangle|^2$.

In quantum theory too the law for time evolution, for example the Schrödinger equation, is *deterministic*. It is the measurement process that introduces *indeterminancy* into quantum theory. In fact, the effect of measurement on the system and apparatus states is *incompatible* with Schrödinger time evolution. While the latter is comparable to *rotations*, the former are like *projections*. These are *the two faces* of quantum theory!

2.1. The demise of the individual!

This immediately makes measurements done on a *single copy* completely without any significance! Firstly, the outcomes are completely random. Secondly, depending on this random outcome, the system state changes *irrevocably*, and with no correlation to the original state one started with. In classical measurements on the other hand, the state is pretty much undisturbed and the statistics of further *repeated* measurements is significant. *The quantum state has no ontological meaning.* This is the *generic* situation.

2.2. The no cloning theorem- an amazing consistency

The no cloning theorem[3] says that an *unknown* state can not be copied. If one could do such a copying, one could have produced an *ensemble* from a single copy. Then ensemble measurements could have determined the state of a single copy! This consistency is almost mystical in quality because no cloning theorem is a property of the reversible, unitary transformations of QM, while the demise of the individual was a consequence of the irreversible and probabilistic nature of measurements!

3. Models for Quantum Measurements (QM)

3.1. The apparatus

The issue of the apparatus in quantum measurements is a rather complex one. Bohr and von Neumann took diametrically opposite stands in this matter.

Bohr, guided deeply by philosophical considerations as well the primary role played by language even in the description of scientific nature, took the stand that the apparatus should be described purely *classically*.

von Neumann, on the other hand went by the premise that if quantum mechanics is a *complete* description, such objects ought to emerge self-consistently from within QM itself. Both views have deep conceptual problems!! It turns out to be very difficult to consistently couple classical and quantum systems (see Sudarshan and Sherry[4]). In fact, the well known Rosenfeld argument for the necessity of quantizing the electromagnetic field is also an example of the type of inconsistencies that result from trying to couple classical and quantum systems. As far as the von Neumann picture, despite important conceptual developments like *decoherence*, a fully satisfactory treatment of the apparatus in quantum measurements, in the opinion of this author, is still lacking.

von Neumann intended his apparatus, though quantum mechanical, to nevertheless mimic a classical apparatus as closely as possible. To appreciate the nuances, let us take a closer look at a classical apparatus. The classical apparatus is characterized by the positions of *pointers* which indicate the values of outcomes. Good resolution means these pointers should not be too thick! In the quantum case, these must correspond to some *Pointer states* of the apparatus which must be mutually *orthogonal*. Different outcomes must correspond to distinct pointer positions, which in QM means that the pointer states must be in one-one correspondence with system states.

Finally, in the classical case pointer positions are unambiguous. This in QM means that for no outcomes should the apparatus state be a *superposition* of pointer states. This last requirement is what turns out to be the most difficult.

3.2. *Macroscopic Objects in Quantum Mechanics*

In fact it even becomes tricky to say when an object behaves *classically* (enough) in QM. Leggett and Garg[5] proposed inequalities to this end and introduced the notion of *Macrorealism*. Unfortunately, these conditions seem only *necessary* but not *sufficient* for the purpose. Significant refinements in their criteria are needed to have a reasonable characterization of what one means by a macroscopic, i.e. classically behaving object in QM. T. S. Mahesh and collaborators[6] in Pune have done beautiful experiments to show how decoherence succeeds in producing states obeying Leggett–Garg inequalities.

3.3. The Dirac-von Neumann model

Now let us consider the first concrete model of quantum measurements, usually ascribed to von Neumann, but we shall include Dirac's name in it as his ideas on the physical interpretation of quantum theory are crucial in its description.[1] According to this, quantum measurements proceed through three distinct stages: (i) The first stage is *initial state preparation* for both the *system* S and the *apparatus* A, (ii) Next, a *measurement interaction* that establishes a one-one correspondence between the states of the system and the apparatus, and finally, (iii) The stage where the measurement is *completed*. Often the second and third stages are confused with each other!!!

The initial state, or the *pre-measurement state* is taken to be:

$$|t_<\rangle_{SA} = |\psi\rangle_S |\phi_0\rangle. \tag{1}$$

von Neumann considered the interaction between S and A to be *impulsive*, described by the unitary transformation:

$$U_M(t_<, t_>) = e^{iS \otimes Q}. \tag{2}$$

The state after this *measurement interaction*, but not after *the measurement* is given by:

$$|t_>\rangle = e^{iS \otimes Q}|\psi\rangle_S|\phi_0\rangle_A = \sum_j c_j|s_j\rangle_S \, e^{is_j Q}|\phi_0\rangle_A. \tag{3}$$

3.4. vN model: the pointer states

Thus, in order for the one-one correspondence established by the measurement interaction to be of the desired type, the apparatus states $e^{is_i Q}|\phi_0\rangle$ should be mutually orthogonal, i.e.

$$\langle\phi_0|e^{i(s_i - s_j)Q}|\phi_0\rangle = 0 \quad i \neq j \tag{4}$$

This poses important restrictions on the initial apparatus state $|\phi_0\rangle$. In particular, $|\phi_0\rangle$ *can not* be an eigenstate of Q. On the other hand, if R is canonically conjugate to Q, i.e. if $[Q, R] = i\hbar$, then $|\phi_0\rangle$ can be an eigenstate of R with, say, eigenvalue r_0.

In the case of Stern–Gerlach experiment, $|\phi_0\rangle$ is a *narrow gaussian* in R. The mutually orthogonal pointer states relevant to the measurement of the observable S with eigenvalues s_i are then the eigenstates $|\phi_i\rangle$ of R with eigenvalues $r_0 + s_i$.

The state after the measurement interaction is, then,

$$|t_>\rangle = \sum_i c_i |s_i\rangle |\phi_i\rangle. \tag{5}$$

This is an *superposition* of all measurement outcomes, not a single outcome! It appears as if one needs a *second* apparatus to look into the first one to see what state it is in. But this does not help as one again ends up with a superposition of the *tripartite* states of the system and the two apparatuses! Wigner even suggested, rather seriously, that *consciousness* has to intervene to break this *infinite regression* named after von Neumann.

Zurek made the very important observation that the system and the apparatus during a quantum measurements do not form a *closed system*, and that one can not ignore the environment. The environment *decorrelates* the various elements of a superposition to *effectively* diagonalize the total density matrix and yield a *mixed* state:

$$\sum_i |c_i|^2 |s_i\rangle\langle s_i| |\phi_i\rangle\langle\phi_i|. \tag{6}$$

3.5. *Issues with decoherence*

It is important to note that a given density matrix will be diagonal only in some very specific bases. Therefore, the decoherence mechanism has to pick such specific *bases* to bring the post-measurement-interaction density matrix to diagonal form displayed above. But the great beauty of quantum mechanics is that it is *basis independent*, in much the same way that special relativity rids physics from being tied to any one particular inertial frame. This aspect of decoherence needs a better understanding. Some quoted *decoherence times* are either too short or too long. *Controllable decoherence* is becoming technologically feasible, so the decoherence picture should be put to stringent *experimental tests*.

A very important question that arises is "Where does the final diagonalization of the density matrix take place?" or, equivalently, "When and where does the measurement process gets completed?" If we were to take the familiar Stern–Gerlach experiment, one could have the impression (incorrect) that the measurement 'takes place' at the magnet producing the inhomogeneous field. It can not be at the magnets, as then multiple Stern–Gerlach of the type shown would be impossible. The multiple Stern–Gerlach arrangement would be accomplished, for example, with four SG-type magnets with their fields and orientations so chosen that the system returns to its original *pure* state. This is what Schwinger called *"putting humpty dumpty together*"! In the particular case of the Stern–Gerlach experiment, the likely seat of decoherence is the screen where the silver atoms strike. This is also the region of the highest density that the atom beam encounters. But here too, carefully done experiments controlling the composition of the screen, should be done.

In summary, in the von Neumann model, the measurement interaction is *impulsive*. Nevertheless, the actual measurement, happens over finite time. As we saw, the likely seat of decoherence is at the screens in the case of the Stern–Gerlach experiment, and it takes finite time for the atoms to traverse the distance from the magnet to the screen. Only eigenvalues are the outcomes of such measurements. Expectation values of properties can be *measured* in a statistical sense. Measuring complete set of properties leads to *tomography*. As a result of the completed measurement, irretrievable damage happens to the pre-measurement state. The latter is not true of the measurement interaction, whose effects can always be *reversed*, as in a multiple Stern–Gerlach set up.

4. Arthurs–Kelly Type Measurements

The famous *Heisenberg microscope*, which has made it to most textbooks on QM, was proposed by Heisenberg around 1927, before the vN model. Through this thought experiment, he formulated the famous *indeterminacy relations*:

$$\delta p \, \delta q \simeq h. \tag{7}$$

There were many lacunae of serious conceptual nature with the original analysis of Heisenberg. As pointed out by Bohr, the Heisenberg analysis had not realised the crucial link between indeterminacy and what was called the *wave–particle duality* then. Nevertheless, Bohr accepted the inevitable indeterminacy as a correct conclusion. The above form should not be confused with the so called *Uncertainty Relations* of the type:

$$\Delta X \, \Delta P \geq \frac{\hbar}{2}. \tag{8}$$

While the latter is a statement on the *measurement statistics* from an *ensemble* of measurements, the former was about the outcome of a *single* experiment. Importantly, the Heisenberg thought experiment was at heart a *joint measurement* of both position and momentum. Heisenberg's reasonings were *semi-classical* at best. In retrospect, it overlooked many important aspects of quantum measurements. In particular, the statistical aspects were totally obscure.

It therefore becomes pertinent to ask how this important work of Heisenberg can be properly understood within modern quantum mechanics. The *Arthurs–Kelly* theory of *simultaneous measurement of canonically conjugate variables*[7] achieves precisely that. They use two *apparatuses*, one for position and one for momentum. The two apparatuses belong to independent quantum systems. Both have *finite resolutions*. This may at first sight seem inevitable as otherwise one may get the impression of being able to simultaneously measure conjugate variables with arbitrary precision. But the situation is more subtle, and one can not measure the conjugate variables in a way that would violate the uncertainty relations.

Even in this case, the outcomes are random, and only *ensemble* measurements make sense. What one obtains are *smeared phase space distributions* Being the outcomes of measurements, there is never any question about the *positivity* of these distributions. Remarkably, the uncertainties are bound by:

$$\Delta P \, \Delta Q \geq 2 \frac{\hbar}{2}. \tag{9}$$

The best that can be achieved i.e. the *minimum uncertainty* is worse by a factor of 2 of what the Heisenberg uncertainty relations allow for. Thus, simultaneous measurement of conjugate variables is *noisier*!

5. Protective Measurements

The *Copenhagen interpretation*, in its broadest sense, would say that measurements in quantum mechanics would irretrievably *disturb* the state of a system. One can then ask whether it is at all possible to design a measurement (which will clearly not be of the von Neumann type) which can yield *full* information about a state *without* disturbing it?

Aharonov and his collaborators[8] have proposed a novel form of measurement which they call *protective measurements*, which in fact answer this question in the affirmative! But the catch is that it works only with a select class of states undergoing select evolution. According to them, one considers *unknown states* which are, however, known to be *non-degenerate eigenstates* of some *unknown* Hamiltonian. Often, a reaction to this is: 'that is not restrictive, every state is an eigenstate of some Hamiltonian'! But the catch is that this Hamiltonian has to act *at all times*. A radical departure from the von Neumann case is that measurement interaction is now *adiabatic*, and one has to consider the limit $T \to \infty$. One may wonder about the meaning of a measurement that takes forever! What this effectively means that the measurement lasts much longer than the longest characterstic timescale governing the system and apparatus. But we shall soon see that the strict $T \to \infty$ limit plays a rather fundamental role in this type of measurements.

Their result is that in this *extreme adiabatic* limit, the system state remains unchanged (*protected*), while the apparatus position shifts by the *expectation value* of the observable in the state. The state remaining unaltered and the $T \to \infty$ limit may be reminiscent of the adiabatic theorem in quantum mechanics, but the results are deeper than that as the state of the apparatus does change in a rather remarkable way.

Therefore, in this type of measurements, expectation values rather than eigenvalues get to be measured directly. Since the state remains unaltered, it can be used for subsequent protective measurements of other observables. This way, the state of a *single copy* could be determined without *disturbing* it. On this basis, Aharonov and his collaborators have repeatedly claimed that protective measurements give ontological significance to wavefunctions. In other words, the wavefunction in QM should be considered to have a reality. There are at least two major problems with these claims. The first is that protective measurements work only on a limited class of wavefunctions, and at best one could claim that for such a class of wavefunctions, one could provide *reality*!

Tabish Qureshi and myself[9] have shown that even in this limited sense, protective measurements fail to deliver the goods. This is where the strictness of the $T \to \infty$ limit becomes all important. In practice T can never be infinite and even a very tiny $\frac{1}{T}$ gives rise to an *orthogonal* correction. While for ensembles this is of negligible consequence, it has a *detrimental* effect on *single state* measurements. This is because the outcomes are totally *random* and there is no protection against the first outcome going wrong. So even with this ingenious scheme there can be no reality for wavefunctions. However, a pragmatic fallout is that one can maintain purity to a high degree and get full information about the state at the same time. This theme has been dealt at length in.[10]

The workings of a protective measurement is shown schematically in Fig. 1, for the case of a Stern–Gerlach experiment. In the standard version using silver atoms from an oven(hence with high velocities and therefore small times of interaction with the magnet), the atoms strike the screen at

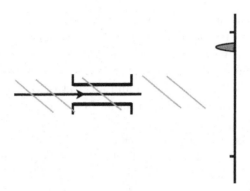

Fig. 1.   Protective Stern–Gerlach measurement.

discretely many points(two if the atoms are spin-$\frac{1}{2}$). The spot where they strike is determined by the eigenvalue of spin. In the protective version (to be done with very cold atoms), they strike the screen at a single spot determined by the expectation value.

6. Weak Measurements

We conclude by discussing a totally different variant of quantum measurements called the *weak measurements*, and a further variant of them which may be called *weak value measurements* .[11] Recall that in the von Neumann type measurements, also called *Projective Measurements*, the initial apparatus state $|\phi_0\rangle$ had to be chosen as an eigenstate of R, or more practically, as a *narrow* wavepacket centred around some value r_0 of R(and therefore a *broad* wavepacket in Q).

In weak measurements, one goes to the opposite extreme: the initial state of the apparatus is taken to be a very broad wavepacket in R. This can be represented as an almost equal superposition over the pointer states $|\phi_i\rangle$ introduced earlier. To be concrete, let us consider the weak version of a SG-measurement. In the vN version, there were only *two* possible outcomes. But now there are many many possible outcomes. If the outcome r_i is much smaller than the width Δ, then all the pointer state wave functions are practically equal, and the system state practically the same as what one started with. Therefore, with a *high probability* the system is *undisturbed* (in much of the literature, this probabilistic aspect is not stressed, resulting in misleading statements). When the outcome is large, depending on its sign, it mimics the two outcomes of a projective measurement.

As in the case of all varieties of quantum measurements, weak measurements can also be performed either on ensembles or on single copies. In the ensemble version, the states undergo very little change and yet information can be obtained as *on the average* the pointer states can undergo measurable changes. The large width necessitates a very, very large ensemble size to control statistical errors.

When done on a single copy, since the state has changed very little (most of the time), it may makes sense to perform *repeated* measurements on it. The precise significance of the ensuing *statistics of sequential measurements* is hard to analyze. It seems unlikely that the state of a single copy can be determined this way. Since weak measurements can be derived within standard quantum mechanics, they are unlikely to violate its basic tenets.

7. Weak Value Measurements

Aharonov, Albert and Vaidman[11] also proposed a variant of the weak measurements whereby the measurement interaction of weak measurements performed on a *pre-selected* state is followed by a *post-selection*, done via a projective measurement. Then the average of the outcomes allows the direct determination of what they called a *weak value*, defined by

$$A_w = \frac{\langle \psi_{post}|A|\psi_{pre}\rangle}{\langle \psi_{post}|\psi_{pre}\rangle}. \tag{10}$$

Such a weak value can be very large, larger than the maximum eigenvalue of the observable A, if the pre and post-selected states are almost orthogonal. In fact, it can even be complex, in which two separate weak measurements are necessary. Lot of confusing and misleading statements have appeared in literature about this weak value.

The information-theoretic aspects of the weak values is yet to be understood fully. A curious property of weak values is that $\langle (A+B)\rangle_w = \langle A\rangle_w + \langle B\rangle_w$.

8. Work in Progress

We are able to show that weak values can be obtained in strong measurements also. Aharonov *et al.* had to restrict the observables to *weak operators*, but we can easily generalize to operators of arbitrary strength. We are studying whether weak values obtained thus are also effective in resolving various *Quantum Paradoxes*.[12] The relative merits of these two ways for getting weak values is being studied. The precise significance of the statistics of sequential weak measurements is also being studied.

Acknowledgment

I would like to thank Professor B.V. Sreekantan for his kind invitation to present these ideas at this meeting. I thank the Abdus Salam International Center for Theoretical Physics (ICTP, Trieste), where this manuscript was finalised, for its hospitality. I acknowledge support from the DST project IR/S2/PU-001/2008.

References

1. Dirac, P.A.M., *Principles of Quantum Mechanics (4th Edition)* (Cambridge University Press).
2. von Neumann, J., *Mathematical Foundations of Quantum Mechanics* (Princeton University Press).
3. Wootters, W. and Zurek, W., *Nature* 299, p.802, 1982; D. Dieks, *Phys. Lett. A* 92, p.271, 1982.
4. Sudarshan, E.C.G. and Sherry, T., *Phys. Rev. D* 20, p.857, 1979.
5. Leggett, A.J. and Garg, A., *Phys. Rev. Lett.* 54, p.859, 1985.
6. Athalye, V., Roy, S.S., and Mahesh, T.S., *Phys. Rev. Lett.* 130402, 2011.
7. Arthurs, E. and Kelly J.L., *On the simultaneous measurement of a pair of conjugate observables*, Bell System Technical Journal, v44, i4, 1965.
8. Aharonov, Y. and Vaidman, L., *Phys. Lett. A* 178, p.38, 1993; Aharonov, Y., Anandan, J. and Vaidman, L., *Phys. Rev. A* 47, p.4616, 1993.
9. Hari Dass, N.D. and Qureshi, T., *Phys. Rev. A* 59, p.2590, 1999.
10. Hari Dass, N.D., *Experiments for realising pragmatic protective measurements* (AIP Conf. Proc. 1384, p.51, 2011).
11. Aharonov, Y., Albert, D.Z. and Vaidman, L., *Phys. Rev. Lett* 60, p.1351, 1988.
12. Aharonov, Y. and Rohrlich, D., *Quantum Paradoxes* (John Wiley and Sons, 2008).

Chapter 3

Predictive Information for Quantum Bio-Systems

Arun Kumar Pati

Quantum Information and Computation Group,
Harish-Chandra Research Institute, Chhatnag Road,
Jhunsi, Allahabad 211 019, India
akpati@hri.res.in

We consider the evolution of a quantum bio-system that interacts with an external environment in a stochastic manner. We ask an important question: when can a bio-system be more predictive to a changing environment? We prove that the non-predictive information for a driven quantum bio-system is lower bounded by the change in the quantum correlation and upper bounded by the entropy production in the system and the environment. We argue that for a system to have more predictive information, it must retain the quantum correlation. This shows that at a fundamental level if a biological system has to be energetically efficient, it must minimize the loss of quantum correlation.

1. Introduction

Any physical system undergoing some dynamical change is capable of doing 'computation'. If the dynamics is governed by classical physics, it is classical computation. Similarly, if the dynamics is governed by quantum physics, we call it quantum computation. All natural phenomena are described by quantum theory and classical physics is an approximation. Probably, natural evolution has found a way to exploit quantum principles that we do not understand yet! Our current understanding is that quantum mechanics is universally valid. Not only are matter and interactions governed by the laws of quantum physics, but many weird features such as the quantum superposition have been observed from micro-scales to fullerenes and entanglement has been detected over long distances.

Any living system, to a zeroth order approximation, is a physical system (with complex chemical reactions going on inside). Since biology is based on chemistry, and chemistry is largely based on quantum theory, one may say that quantum mechanics may govern the activities of living systems. It is possible that life has evolved by exploiting some quantum phenomena, including coherence and other weird features. Life has evolved in this planet over a long time and our understanding of quantum mechanics is only about a hundred years old. One of the most pressing questions in recent times is: Do we need quantum superposition for some biological functions? Or are there other kinds of quantumness that may be exploited for living systems to be energetically efficient[1]? The kind of quantum correlation that is responsible for predictive information may exist even at room temperature.

The notion of predictive information has found many applications, including in biological systems. This is so, because, a useful model of a living system must capture predictive information.[2–5] Can biological systems exploit quantum theory to have more predictive information?

Recently, in a classical probabilistic model, it was shown how a system can perform computation by responding to the changing environment. This idea also applies to the living system as a computing system where future expectations are based on their prior experiences. We know that biological computation is fundamentally a nonequilibrium process. In this process, typically large energies liberated drive the operative degrees of freedom of biological machines away from equilibrium averages. It has been shown that the instantaneous nonpredictive information is directly related to the energy dissipated when an external driving signal undergoes stochastic change. This shows that if a classical system makes effective use of information and operates efficiently in the thermodynamic sense, i.e., if the system tends to keep memory of its environment and also operates with maximal energy efficiency, then it has to be predictive.[6]

In this work, we generalize the notion of predictive information for a quantum bio-system driven away from equilibrium. We prove that the nonpredictive information for the driven quantum system is lower bounded by the change in the quantum correlation and upper bounded by the entropy production in the system and the environment. We argue that for a system to have more predictive information, it must retain the quantum correlation. Also, our result shows that if a system has to operate efficiently with minimal entropy production, then it should contain more predictive information. This latter, conforms to the classical probabilistic model. This shows that if a system at a fundamental level has to be efficient, then it must contain maximal predictive information and the quantum correlation is responsible for that. The kind of quantum correlation that we will discuss in this paper is something that goes beyond quantum entanglement. These correlations can be present for quantum states for which there is no entanglement present in the system.

Suppose a quantum system is driven by an external environment. The system in question can maintain an estimate of its uncertainty about future observations and can also estimate the information in each observation about an event that is to be observed in future, given all the observational information of the past. The instantaneous predictive information is a measure of how well the system can adapt to a changing environment at a particular instant of time. It is known that the extraction of predictive information in biological signal processing occurs efficiently. In the sequel, we show that indeed, for driven quantum systems away from equilibrium the non-predictive information is related to both the entropy production and the change in the quantum correlation.

2. Predictive Information

Consider the system A, in a thermodynamic equilibrium at time $t = 0$. The system is driven out of equilibrium when the environment B undergoes a stochastic change. At any time t, the state of the system AB is given by $\rho_{A_t B_t}$, with $\rho_{A_t} = \text{Tr}_B[\rho_{A_t B_t}]$ and $\rho_{B_t} = \text{Tr}_A[\rho_{A_t B_t}]$. The dynamics is given by a sequence of unitary evolutions for AB and noise acting on B. In a realistic scenario this actually happens. The system tries to interact with its surroundings and learns about it. In the meantime, the environment can undergo some change (due to action of some channel). Then, the system interacts again with the environment and the process continues.

This can be represented as $\rho_{AB} \rightarrow U_0(\rho_{AB})U_0^\dagger = \rho_{A_0 B_0}$. Then, $\rho_{A_0 B_0} \rightarrow (I \otimes \mathcal{E}_1)\rho_{A_0 B_0} = \rho_{A_0 B_1}$ and $\rho_{A_0 B_1} \rightarrow U_1(\rho_{A_0 B_1})U_1^\dagger = \rho_{A_1 B_1}$. The process starts from $t = 0$ and is repeated till $t = T$.

The dynamics can be compactly specified by these equations:

$$\rho_{A_{t-1} B_t} = (I \otimes \mathcal{E}_t)\rho_{A_{t-1} B_{t-1}}$$
$$\rho_{A_t B_t} = U_t(\rho_{A_{t-1} B_t})U^\dagger$$
$$\rho_{A_t B_{t+1}} = (I \otimes \mathcal{E}_{t+1})\rho_{A_t B_t} \tag{1}$$

We have allowed unitary interactions between A and B. However, one may allow a general quantum operation between A and B also. Since any general quantum evolution can be realized by attaching an ancilla to AB and allowing these to evolve unitarily, we restrict ourselves to unitary interactions. In this model, the state of the system A contains some memory of the environment B. Like in the classical scenario, the past environmental effects are mapped to the current state of the system at time t. As the systems A and B evolve, they get correlated and through the correlations, it may contain a prediction of the future. This can be thought of also as a process of learning (the system tries to predict the future, based on the observation of past behavior).

The system's state and the environment share some information at t. The instantaneous memory is the mutual information for state $\rho_{A_t B_t}$ defined as:

$$
\begin{aligned}
I_{memory}(t) := I(\rho_{A_t B_t}) &= S(\rho_{A_t}) + S(\rho_{B_t}) - S(\rho_{A_t B_t}), \\
&= S(\rho_{A_t}) - S(A_t|B_t),
\end{aligned}
\tag{2}
$$

where $S(\sigma) = -\operatorname{Tr}(\sigma \log_2 \sigma)$ is the von Neumann entropy of a quantum state σ and $S(A_t|B_t) = S(\rho_{A_t B_t}) - S(\rho_{B_t})$.

In our model, the dynamics maps $\rho_{A_{t-1} B_t}$ to $\rho_{A_t B_t}$ and this represents the prediction for $\rho_{B_{t+1}}$ in future when B undergoes a general evolution, given the state ρ_{A_t}. A useful notion in this context is the instantaneous predictive information, defined as:

$$
\begin{aligned}
I_{pred}(t) := I(\rho_{A_t B_{t+1}}) &= S(\rho_{A_t}) + S(\rho_{B_{t+1}}) - S(\rho_{A_t B_{t+1}}) \\
&= S(\rho_{A_t}) - S(A_t|B_{t+1}).
\end{aligned}
\tag{3}
$$

This represents information that system A shares at time t with the system B at time $t+1$. Therefore, the instantaneous non-predictive information is the difference $I_{memory}(t) - I_{pred}(t)$.

Since $I_{pred}(t) = I \otimes \mathcal{E}_{t+1} I(\rho_{A_t B_t})$, this cannot increase under a general quantum channel.[11] Hence, the non-predictive information is always positive. In the classical context the non-predictive information is proportional to average work dissipated, thus leading to the result that the unwarranted retention of past memory is fundamentally equivalent to energy inefficiency.

3. Predictive Information, Entropy and Quantum Correlation

We show here that $I_{nonpred}(t) - I_{memory}(t) - I_{pred}(t)$ is lower bounded by the quantum correlation change and upper bounded by the total entropy change of the system AB (and the supra-environment E). To be precise, we prove that:

$$\Delta Q_t \leq I_{nonpred}(t) \leq \Delta S_t, \tag{4}$$

where the change in the quantum correlation $\Delta Q_t = Q(\rho_{A_t B_t}) - Q(\rho_{A_t B_{t+1}})$ and $\Delta S_t = \Delta S_{A_t B_t} + \Delta S_{E_t}$.

The supra-environment can be thought of as the rest of the universe, to which we do no have direct access. Since the quantum channel acting on the environment B can be thought of as a unitary evolution on B and another subsystem E, we call that as supra-environment.

To prove this, we will show that the instantaneous memory and predictive information have a quantum and classical part, i.e., $I_{memory}(t) = Q(\rho_{A_t B_t}) + C(\rho_{A_t B_{t+1}})$ and similarly, $I_{pred}(t) = Q(\rho_{A_t B_{t+1}}) + C(\rho_{A_t B_{t+1}})$.

First, we show that $I_{memory}(t)$ can be split into two parts: one the quantum part and the other, the classical[9,10] part. The quantum part represents the instantaneous quantum correlation, which is called as the discord.[10] For any bipartite density operator $\rho_{A_t B_t}$, at time t, we can define the quantum memory as the difference between the total memory and the classical memory. The classical mutual information is given by[9]:

$$C(\rho_{A_t B_t}) = S(\rho_{A_t}) - \min_{\{\Pi_i^B\}} \sum_i p_i S(\rho_{A_t|i}), \tag{5}$$

where the conditional entropy upon measurement is $S(\rho_{A_t|B_t}) = \min_{\{\Pi_i^B\}} \sum_i p_i S(\rho_{A_t|i})$, the minimization being over all projective measurements performed on subsystem B.

The probability for obtaining i outcome is $p_i = \mathrm{Tr}_{AB}[(I_A \otimes \Pi_i^B)\rho_{A_t B_t}(I_A \otimes \Pi_i^B)]$, and the corresponding post-measurement state for the subsystem A is given by $\rho_{A_t|i} = \frac{1}{p_i}\mathrm{Tr}_B[(I_A \otimes \Pi_i^B)\rho_{A_t B_t}(I_A \otimes \Pi_i^B)]$, where I_A is the identity operator on $\mathcal{H}_A$. The quantity $C(\rho_{A_t B_t})$ is the classical memory and can be interpreted as information gain about subsystem A_t as a result of a measurement on the subsystem B_t. The difference $[I(\rho_{A_t B_t}) - C(\rho_{A_t B_t})]$ is a measure of quantum correlations, which represents the quantum part of the instantaneous memory. Since the quantum mutual information is never lower than the classical memory, the quantum correlation is always positive.

Therefore, the quantum correlation is given by[10]:

$$Q(\rho_{A_t B_t}) = \min_{\{\Pi_i^B\}} \sum_i p_i S(\rho_{A_t|i}) - S(A_t|B_t), \tag{6}$$

where $S(A_t|B_t) = S(\rho_{A_tB_t}) - S(\rho_{B_t})$ is the conditional entropy. Here, $S(\rho_{A_t|B_t}) = \min_{\{\Pi_i^B\}} \sum_i p_i S(\rho_{A_t|i})$ is always a positive quantity and can be thought of as a measure of the uncertainty left on average, about A given that B has been measured. It may be noted that for driven classical systems both $S(\rho_{A_t|B_t})$ and $S(A_t|B_t)$ are the same and hence no quantum part exists for the instantaneous memory.

However, for quantum systems the former can be more than the latter, leading to a nonzero quantum memory as measured by the quantum correlation. Using the same reasoning we can write $I_{pred}(t) = Q(\rho_{A_tB_{t+1}}) + C(\rho_{A_tB_{t+1}})$.

Therefore, the nonpredictive information can be expressed as $I_{nonpred}(t) = \Delta Q_t + \Delta C_t$, where $\Delta Q_t = [Q(\rho_{A_tB_t}) - Q(\rho_{A_tB_{t+1}})]$ is the change in the quantum memory and $\Delta C_t = [C(\rho_{A_tB_t}) - C(\rho_{A_tB_{t+1}})]$ is the change in the classical memory. As a classical correlation cannot increase under general evolution,[9] $I_{nonpred}(t) \geq \Delta Q(t)$. This shows that to minimise the nonpredictive information, we should minimise the change in the quantum correlation. Thus, physical processes which have maximal predictive information are those that maintain the quantum correlation.

To prove the upper bound, consider the general evolution acting on B_t which results $\rho_{A_tB_{t+1}} = I \otimes \mathcal{E}_{t+1}(\rho_{A_tB_t})$. This can be regarded as a unitary evolution by bringing a supra-environment E, i.e., $I \otimes \mathcal{E}_{t+1}(\rho_{A_tB_t}) = \mathrm{Tr}_E[U_{BE_{t+1}}(\rho_{A_tB_t}) \otimes \rho_{E_t})U_{BE_{t+1}}^\dagger]$.

We can write the nonpredictive information as:

$$I_{nonpred}(t) = \Delta S_t + [S(\rho_{B_t}) + S(\rho_{E_t})] - [S(\rho_{B_{t+1}}) + S(\rho_{E_{t+1}})], \qquad (7)$$

where $\Delta S_t = \Delta S_{A_tB_t} + \Delta S_{E_t}$ is the total change in the entropy of the system AB and the supra-environment E. We can define initial and final total entropies as sum of the entropies of the bipartite quantum system AB and the environment E before and after the interaction, i.e., as $S_t = S(\rho_{A_tB_t}) + S(\rho_{E_t})$ and $S'_t = S(\rho_{A_tB_{t+1}}) + S(\rho_{E_{t+1}})$.

The change in total entropy is defined as $\Delta S_t = S'_t - S_t = \Delta S_{A_tB_t} + \Delta S_{E_t}$. The change in the total entropy is always positive which follows from the subadditivity of entropy and the unitarity. Thus, we have:

$$I_{nonpred}(t) \leq \Delta S_t + S(\rho_{B_tE_t}) - S(\rho_{B_{t+1}E_{t+1}}). \qquad (8)$$

We may note that at time t, B and E are in the product state, hence $S(\rho_{B_t}) + S(\rho_{E_t}) = S(\rho_{B_tE_t})$. Using the subadditivity of the von Neumann entropy, we have $S(\rho_{B_{t+1}}) + S(\rho_{E_{t+1}}) \geq S(\rho_{B_{t+1}E_{t+1}})$. Since the subsystems

BE evolve unitarily we have $S(\rho_{B_t E_t}) = S(\rho_{B_{t+1} E_{t+1}})$. Therefore,

$$I_{nonpred}(t) \leq \Delta S_t. \tag{9}$$

Thus, the amount of non-predictive information can never exceed that of the total instantaneous entropy change. *This shows that if a system with memory has to be predictive, then it has to minimize entropy production.*

4. Conclusions

Most of the natural systems operating away from equilibrium tend to be predictive in order to be thermodynamically efficient. The nonpredictive information quantifies the ineffectiveness of the system in coping with a changing environment. In this work, we have generalized the notion of predictive information for driven quantum system and have shown that the nonpredictive information is upper bounded by the total entropy change and lower bounded by the quantum correlation change of the combined system and environment. Earlier, it has been shown in the classical context that the non-predictive information is related to dissipation. Our result shows that to minimise the entropy production, the system has to be predictive. Furthermore, for a system to be predictive, it should retain the quantum correlation. These results can be applied to biological systems, quantum information systems, and many other physical processes that are driven by the external environment.

This also shows that the quantum correlation plays a new role: it helps the system to be predictive. Since quantum correlations may exist even at room temperature, this may provide new insights for other biological processes. Our result may provide deeper insights on the role of predictive information in driven quantum systems, living organisms, quantum systems far from equilibrium and many more.

References

1. D. Abbott, P. Davies and A. K. Pati (eds.), *Quantum Aspects of Life* (Imperial College Press, London, 2008).
2. H. Jeffreys, *Theory of Probability*, Oxford University Press, New York, 3rd edition (1998).
3. W. Bialek, in *Physics of Bio-Molecules and Cells*, Proceedings of the Les Houches Summer School, Session LXXV, Springer-Verlag, Berlin (2001), p. 485–577.
4. S. Still, *Europhys. Lett.* **85**, 28005 (2009).

5. S. Still, J. P. Crutchfield, and C. Ellison, *Chaos* **20**, 037111 (2010).

6. S. Still, D.A. Sivak, A.J. Bell & G.E. Crooks, *Phys. Rev. Lett.* **109**, 120604 (2012).

7. L. Henderson and V. Vedral, *J. Phys. A* **34**, 6899 (2001).

8. H. Ollivier and W. H. Zurek, *Phys. Rev. Lett.* **88**, 017901 (2001).

9. L. Henderson and V. Vedral, *J. Phys. A* **34**, 6899 (2001).

10. H. Ollivier and W. H. Zurek, *Phys. Rev. Lett.* **88**, 017901 (2001).

11. M. H. Partovi, *Phys. Lett. A* **137**, 440 (1989).

Chapter 4

Quantum Effects in Biological Systems

Sisir Roy[*]

*Physics and Applied Mathematics Unit, Indian Statistical Institute,
203 B.T. Road, Kolkata 700 108, India*

sisir.sisirroy@gmail.com

The debates about the trivial and non-trivial effects in biological systems have drawn much attention during the last decade or so. What
might these non-trivial sorts of quantum effects be? There is no consensus so far among the physicists and biologists regarding the meaning
of "non-trivial quantum effects". However, there is no doubt about the
implications of the challenging research into quantum effects relevant to
biology such as coherent excitations of biomolecules and photosynthesis,
quantum tunneling of protons, van der Waals forces, ultrafast dynamics
through conical intersections, and phonon-assisted electron tunneling as
the basis for our sense of smell, environment assisted transport of ions
and entanglement in ion channels, role of quantum vacuum in consciousness. Several authors have discussed the non-trivial quantum effects and
classified them into four broad categories: (a) Quantum life principle;
(b) Quantum computing in the brain; (c) Quantum computing in genetics; and (d) Quantum consciousness. First, I will review the above
developments. I will then discuss in detail the ion transport in the ion
channel and the relevance of quantum theory in brain function. The ion
transport in the ion channel plays a key role in information processing
by the brain.

1. Introduction

Every object in this universe is made of electrons, protons, quarks etc., and
these subatomic particles are governed by the laws of quantum physics. So
the functional processes in a living organism may be explained by the laws

---

[*]Section 5: "Decoherence time and Quantum Coherence" has been contributed jointly by
Sisir Roy and Samyadeb Bhattacharya, Physics and Applied Mathematics Unit, Indian
Statistical Institute, Kolkata, India.

of quantum physics. We are not much concerned with this fact in quantum biology. One of the biggest questions in biology is whether the processes of life are able to exploit quantum effects to improve their lot. If nature has not found a way to exploit quantum mechanics, an equally important question is: why not? Is it merely an oversight on the part of evolution or is there some other deeper reason why evolution cannot exploit quantum mechanics?

Till recently, basic quantum phenomena were believed to occur not beyond the level of a few subatomic particles so that at larger scales these were absent. Yet there were still some that believed that quantum theory would explain the basis of life as we see it.

Recently Lambert et al.[1] made an overview of the two main candidates for biological systems which may harness such functional quantum effects: photosynthesis and magnetoreception. Before going into details let us discuss some of the important concepts in quantum theory called "quantum weirdness".

2. Quantum Weirdness

The following concepts are extensively discussed in this context: superposition principle, nonlocality, quantum coherence and entanglement.

- Superposition: While certain physical states are rendered virtually impossible and cannot coexist in classical physics, quantum mechanics allows a superposition of various quantum states, each with a finite probability of existence.
- Concept of Nonlocality: Einstein–Podolsky–Rosen (EPR) suggested an experiment which gives rise to the debate on the "incompleteness" of quantum mechanics (QM). The concept of nonlocality was introduced in QM. This is a metaphysical concept. Nonlocality implies the presence of action at a distance, with distinct and distant parts of the universe having an immediate connection.
- Quantum Entanglement: Quantum entanglement is a strange and non-intuitive aspect of the quantum theory of matter, which has puzzled and intrigued physicists since the earliest days of the quantum theory. Quantum entanglement represents the extent to which measurement of one part of a system affects the state of another; for example, measurement of one electron influences the state of another that may be far away.

- Quantum Coherence: In quantum mechanics, unlike the classical case, coherence can be displayed by particles due to interference of probability amplitudes. One typical example is, of course, the oscillating net amplitude seen on a screen after allowing a stream of particles to pass through either of two holes in a barrier in "2-slit experiment". This is a resultant of the addition of the complex amplitudes associated with either of the two individual possible paths, allowing for a cross-term to appear in the net amplitude. Neglecting the cross-term reduces the result to that (non-oscillating diffraction pattern) seem in the classical case.[9]
- Classical Coherence: In classical wave optics, one can get interference of waves and define coherence too. For a harmonic oscillator, calculation of the position-position time correlation function shows that the system has coherent behavior at its frequency of vibration.[9] How does one distinguish between coherence phenomena that are of a classical origin and those that are intrinsically quantum mechanical? Coherence effects may be of quantum or classical origin and it is not always so obvious which it is. To make matters even more ambiguous, sometimes, whether the observed coherence stems from a quantum mechanical or classical origin, depends upon what is being observed!
- Quantum Weirdness in Biology: At physiological temperatures, reactions involving proton-transfer or electron-transfer steps at the cellular level are present. Such reactions would invariably include quantum effects. Yet, their consideration is deemed secondary: the reasoning being that such transfers occur only after the classically-described behaviour of the more massive protein molecules which have already largely influenced the reaction energetics. Thus, quantum dynamics are usually ignored in simulation studies of biolecular behavior.[10]

3. Quantum Coherence in Biology

We will concentrate our discussions on the issue of quantum coherence in the following areas of biology.

- Photosynthesis: The evidence for quantum coherence is considered to be exceptionally strong.
- Sense of Smell: The conventional postulate for the mechanism of

smell is a lock and key model, in which different types of odorant molecules bind to different types of olfactory sensors. However, the nose actually behaves like a vibrational spectrometer. The only known mechanism that can explain this vibrational sensitivity of the sense of smell is quantum mechanical.

- The Avian Compass: Some birds, such as homing pigeons, possess a small piece of magnetite in their beaks which functions as a compass, allowing them to tell North from South.

3.1. Quantum Entanglement and Photosynthesis

The Fenna-Matthews-Olson (FMO) complex (a light harvesting complex) can be seen as a multiarm interferometer where each arm of the interferometer corresponds to a site of FMO complex. The propagation of an exciton through the FMO complex is analogous to propagation of a photon through an interferometer. A single photon propagating through the interferometer produces entanglement between the arms and so the propagation of excitation may also produce entanglement between the sites of the FMO complex. However, there are some open issues which need to be analyzed carefully before making any conclusive remark. First, the entanglement between the arms of the interferometer occur depending on the initial state concerning photon. In the case of photosysnthesis in which the entire light harvesting complex (FMO complex) is illuminated by weak classical light, not by a single photon, the entanglement may therefore occur in the FMO complex.

3.2. Quantum Physics and Olfaction

Smell is one of our five basic senses, but the key steps in the mechanism of smell (olfaction) remain unknown. We know that particles move from the source of smell through the air to our nasal membranes by the process of diffusion, and we are very familiar with the neurological pathways that happen after the odour has been detected, but the happenings in between are still in debate.

A standard theory of olfaction is what is termed as the Shape Theory. This requires molecules and receptors to have particular shapes, so that only those odorant shapes that lock into corresponding receptors of appropriate shapes are detected with a particular smell. This ignores the fact that similarly shaped molecules may have very distinct smells and conversely, very different molecular shapes may have similar smells. This has caused the proposal of another theory called Vibration Theory, which deals with the specifics of molecular vibrations. This deals with quantum-mechanical effects such as electrom tunneling through potential barriers. If true, quantum mechanics is directly responsible for detection of smells.

3.3. *Navigating Birds and Magnetoreceptor*

Bird navigation is a complex enterprise, requiring birds to make repeated and varying orientation decisions based on directional and positional information. Birds are aided by multiple physiological compass systems, among them a physiological magnetic compass. The existence of a magnetic compass was discovered in orientation experiments with birds in cages by Wolfgang Wiltschko in the late 1960s. In the 1970s, studies indicated that weak magnetic fields can influence chemical processes involving photoactivated radical pair intermediates, i.e., a transient pair of molecules with an unpaired electron spin each. The underlying mechanism was shown to be based on the effects of magnetic fields on the electron spin evolution in each of the radical pairs and investigation of such effects opened the now mature field of spin chemistry. It was Klaus Schulten who first suggested that this radical pair mechanism might operate in the compass of migratory birds.

4. Ion Channel and Information Propagation in Brain

The brain's treatment of information can be characterized by the time scales over which the various aspects are treated. At smaller time scales, the electrical activity of neurons is the medium for the transmission and processing of information. At larger time scales, information storage is codified through the strength of the inter-neuronal synaptic connections. Describing neuronal behavior requires varied approaches. While membrane proteins in neurons known as ion channels control ion-flow (and hence electrical activity) through an individual stochastic process of opening and closing, the macroscopic behavior of these can be approximated as deterministic. While the function of man-made computers as well as biological brains are both based on electrical signals, integrating the behavior of a solid-state device with that of living, biological tissue with polar liquids is a complex challenge. Yet we are today in a position to actively pursue this more comprehensively.[11] Conceptually three significant functional domains of all ion channels are:

- Ion Conducting Pore: An aqueous pathway for ions with a narrow selectivity filter that distinguishes among the ions that do go through and the ions that do not.
- Gates: A part of the channel that can open and close the conducting pore
- Sensors: Detectors of stimuli that respond to electrical potential changes or chemical signals. The sensors couple to the channel gates to control the probability that they open or close.

Scientists try to explain the dynamics of ion channel using the two approaches within classical physics:

- Molecular dynamics
- Brownian dynamics

However, the present author along with his collaborators have shown that quantum mechanical description is needed to explain the dynamics of K^+ ion channel and it raises a lot of interesting issues such as decoherence time, etc.

5. Decoherence Time and Quantum Coherence

The approaches towards dynamics of ion transport in protein membranes (ion channels) are generally considered to be classical. Recently, one of the present authors showed that[2] the dynamics of K-ion channel can be explained using nonlinear Schrödinger equation which is compatible with the results of MacKinnon's experimental observation.[13] The two K-ions may form an entangled state within a selectivity filter during a finite period of time. The temperature within the channel is generally considered to be high enough to destroy the coherence within a very short period. Moreover, molecular modes of the protein environment induce dynamical decoherence, which destroys the quantum mechanical superposition of states in a very short period of time. So a quantum mechanical approach in this area of research was mostly speculative and far from experimental realization. But in recent times, some experimental demonstration of the presence of quantum coherence in the process of photosynthetic energy transfer[1,2] lead us to reconsider the theoretical approach and understanding.

Here we are concerned with the question that whether and under what conditions a sustainable quantum superposition is achievable in the process of transfer of ions through biological channels. Now it is quite a practical argument that the quantum state of the traversing ion is strongly coupled with the molecular vibrational modes of the protein environment and hence fast decoherence is almost absolutely unavoidable. But it is to be noted that the traversal time of the ion through the membrane is also quite small and it is the ratio of the time scale of decoherence with this traversal time that plays an important role in understanding the maintenance of coherence. If the decoherence time is larger than the traversal time of the ion, then the quantum superposition of the ionic states is sustainable enough for the traversing entity within the period of ionic transfer. Here we also need to consider the effect of temperature in estimating the decoherence time.

With the recent developments of quantum thermodynamics,[11] a new concept of temperature (known as spectral temperature) for microstates in the non-equilibrium condition has been proposed, instead of the usual concept of thermodynamic temperature. Since ion transport through protein membrane is essentially a non-equilibrium phenomenon, we propose that the spectral temperature plays an important role in understanding the coherence in the channel dynamics.

From numerous definitions of tunneling time,[3] we take dwell time as the necessary time taken by the tunneling entity to traverse the barrier. In a previous work, we have calculated the weak value of dwell time as

$$\tau_D = \frac{1}{\gamma} \coth\left(\frac{\gamma \tau_M}{2}\right) \tag{1}$$

where τ_M is the measurement time. Here this measurement time can be interpreted as the time interval between gate opening and closing of the ion channel. For a non-dispersive barrier, the dwell time can be interpreted as the time interval between the energy storage and release in the barrier region.[8] The protein membrane embedding the tunneling region can be interpreted as an array of molecules with certain vibration modes, into which the tunneling entity is losing its energy.

From the master equation for open quantum system, the decoherence time can be written as

$$\tau_{dec} = \frac{\hbar^2}{2m\gamma K_B T(x - x')^2}. \tag{2}$$

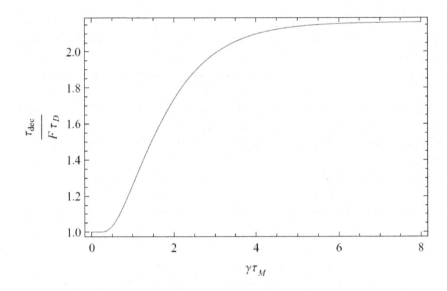

Fig. 1. τ_{dec}/τ_D vs $\gamma\tau_M$. Here also we keep γ as a constant and basically study the variation with increasing τ_M. Here $F = \frac{2\hbar}{w}\sqrt{\frac{2}{m\epsilon_0}}$. As with increment of τ_M, the process becomes quasi-static, we see that the ratio of the two time scales also reach a stable value.

Here, as a model system, we consider a double well potential of the form

$$V(x) = \frac{1}{2}m\omega^2 x^2 \left[\left(\frac{x}{a}\right)^2 - 14\left(\frac{x}{a}\right) + 45\right]. \tag{3}$$

The bistable potential is useful in practical situations, because the ionic transfer can be interpreted as tunneling between the two stability regions.

Since ionic transfer in a protein membrane is a dynamical process subjected to energy exchange with the protein environment, the usual concept of thermodynamical temperature in an equilibrium situation will not suffice. This is why we introduce the concept of spectral temperature originally formulated by Gemmer et al.[11] It is defined as a function of microstates, in order to include non-equilibrium situations.

Here we only consider the ground states of both the potential wells to interpret it as a two-state system. Using Eqs. (1) and (2) and the concept of spectral temperature for two state systems for the double well potential given by Eq. (3), we find that the ratio of the decoherence and dwell time can be expressed as

$$\frac{\tau_{dec}}{\tau_D} = \frac{2\hbar}{w}\sqrt{\frac{2}{m\epsilon_0}}\coth\left[\frac{1}{2}\coth\left(\frac{\gamma\tau_M}{2}\right)\right] \tag{4}$$

where $\epsilon_0 = E_1 - E_0$ is the asymmetry energy of the potential and $w = \frac{15a}{2}$ is the separation length between the wells.

6. Conclusions

Based on the above analysis, we suggest that the ion selectivity filter may exhibit quantum coherence which can play a crucial role in the process of selectivity and conduction of specific ions in biological membranes. For the time scales shorter than that of decoherence time, quantum coherence can be expected to be sustained, and to have vital importance in the dynamics, despite the presence of an interactive protein environment. Our analysis shows that for a sort of quasi-static situation, where the gate opening and closing mechanism is slower than the relaxation (dissipation) time scale, the decoherence-dwell time ratio reaches a static value and can also be greater than unity depending on the mass, energy and length parameters. In such situations, coherent phenomena like entanglement can be of vital importance in understanding the mechanism of selectivity and transport.

7. Acknowledgement

Section 5: "Decoherence Time and Quantum Coherence" has been contributed jointly by Sisir Roy and Samyadeb Bhattacharya, Physics and Applied Mathematics Unit, Indian Statistical Institute, Kolkata, India.

References

1. Lambert et al., Functional quantum biology in photosynthesis and magnetoreception, arXiv:1205.0883V1.
2. Sisir Roy and Rodolfo Llinás, *C. R. Biologies* **332**, 517 (2009).
3. D. Doyle, R. MacKinnon et al., *Science* **280**, 69 (1998).

4. G.S. Engel et al., *Nature* **446**, 782 (2007).
5. I.P. Mercer et al., *Phys. Rev. Lett.* **102**, 057402 (2009).
6. J. Gemmer, M. Michel and G. Mahler, *Quantum Thermodynamics*, Vol. LNP657 (Springer, Heidelberg, Berlin, 2004).
7. E.H. Hauge and J.A. Stoveng, *Rev. Mod. Phys.* **61**, 917 (1989).
8. H.G. Winful, *New J. Phys.* **8**, 101 (2006).
9. W.H. Miller, *J. Chem. Phys.* **136**, 210901 (2012).
10. P.G. Wolynes, *Proc. Nat. Acad. Sci. USA* **106**, 17247 (2009).
11. P. Fromherz, *Physica E* **16**, 24 (2003).

Chapter 5

Instabilities in sensory processes

J. Balakrishnan

Complex Systems Group, School of Natural Sciences & Engineering,
National Institute of Advanced Studies,
Indian Institute of Science Campus, Bangalore 560012, India
janaki05@gmail.com

In any organism there are different kinds of sensory receptors for detecting the various, distinct stimuli through which its external environment may impinge upon it. These receptors convey these stimuli in different ways to an organism's information processing region enabling it to distinctly perceive the varied sensations and to respond to them. The behavior of cells and their response to stimuli may be captured through simple mathematical models employing regulatory feedback mechanisms. We argue that the sensory processes such as olfaction function optimally by operating in the close proximity of dynamical instabilities. In the case of coupled neurons, we point out that random disturbances and fluctuations can move their operating point close to certain dynamical instabilities triggering synchronous activity.

1. Introduction

The living being distinguishes itself from the inanimate through its ability to keep running within itself some self-sustained processes and signaling mechanisms. It performs certain vital functions to sustain itself. The underlying physical and chemical processes may be sought to be described qualitatively and quantitatively through the framework of appropriate mathematical models, using tools and techniques of dynamical systems theory.

Here, we discuss how techniques of dynamical systems theory and concepts of physics can be put to fruitful use for understanding aspects of cognitive behavior of organisms at the elemental, cellular scale.

The dynamical behavior of a system may, broadly speaking, lie in the oscillatory regime, or in the chaotic regime. Bifurcations or dynamical

37

instabilities determine the future turn of events of the system. In sensory processes too, instabilities play a key role in bringing about:

(1) stimulus detection,
(2) spontaneous activity of sensory cells,
(3) neuronal firing,
(4) changes in signaling patterns, and
(5) synchronous activity.

Our focus here is on oscillatory instabilities since they are endowed with several interesting features.

In undriven systems, oscillatory dynamics can be brought about when the operating point of the system approaches the vicinity of a bifurcation point as a parameter varies. This can happen in several ways, the best known being the Andronov-Hopf bifurcation, wherein the eigenvalues of the linearized system become purely imaginary at the critical point and a limit cycle is born. Another type of bifurcation can occur when the flow generated by the differential equation describing the system is on a circle, due to destruction of fixed points as they coalesce when an unstable manifold is present; the saddle-node bifurcation on an invariant circle is a well known prototypical example.

A nonlinear oscillator poised in the close proximity of a Hopf bifurcation exhibits several properties which can be of practical interest:

First, such an oscillator displays extreme sensitivity to changes in its parameter space.
Second, it has a large dynamical range.
Lastly, a Hopf oscillator can show spontaneous oscillatory behavior, given satisfactory conditions.

The first property actually implies that the system has the ability to greatly amplify input signals of very small strengths. If f denotes the signal strength, then at the critical point, a system showing a Hopf bifurcation displays a response $\sim f^{1/3}$ so that its gain varies as $|f|^{-2/3}$. Thus, small inputs are translated with huge gains. This useful feature, combined with the second property, makes a Hopf oscillator an excellent signal detector. Therefore we have the interesting result that when a nonlinear system is tuned to operate very close to its Hopf bifurcation point, it would show optimal performance as a good signal detector.

The premise that the sound detection mechanism in an organism is powered by dynamics of a Hopf bifurcation has, in recent times, gathered strong observational evidence.[1-3] So has Nature, with a sense of economy, used the same principle in the design of other sensory detection systems in organisms? There is, yet, no proven basis for the existence of a generic mechanism operating near a dynamical instability, for the detection of other modalities like taste, smell, vision, etc., all of which are detected by chemosensory receptors. Nevertheless, knowing the general features and characteristics displayed by chemoreceptors, we may hypothesize the existence of a broadly generic, similar principle driving their individual mechanisms to account for their ability to optimally detect stimuli.

It has so far been believed that olfactory receptors sense structural motifs of the odorant molecules (odotopes) which bind to them, enabling them to detect and distinguish between different smells. This theory is not very satisfactory as it is unable to explain why certain dissimilar molecules can smell similar or why certain molecules with identical shapes can smell very different. On the other hand, in the literature, there has been revival of an old hypothesis,[8] backed by recent experiments,[9] that olfaction is a spectral sense like hearing and vision. According to this, olfactory receptors detect odorants by their molecular vibration through a quantum mechanical process (phonon-assisted inelastic electron tunneling) when an appropriate odorant molecule or "odotope" fits in the receptor. This theory, based on the vibrational spectrum of the odorants, explains why molecules containing different isotopes can smell differently and why molecules with different structures can smell similar. Dissimilar molecules which smell similar have similar vibrational spectra.

In what range then would the phonon frequencies lie? Odorant molecules are typically of low molecular mass, say a few hundred Daltons. A rough, back of the envelope calculation may be performed to estimate the phonon frequencies involved. Assuming that the values of the elastic spring constants of the inter-atomic forces in the odorant molecules lie in the range $1 - 54$ Nm, we find that the phonon frequencies lie in the range 10^{13} and 10^{14} Hz (far infrared).

We recall that the signal detection and transduction process in a sensory cell involves situations that are far from equilibrium, where even single molecule events are important. The chemosensory signal detection mechanism involves ligand-receptor binding mediated by enzymes and catalytic reactions. At that scale, we may expect fluctuations in the concentrations of participating molecules to significantly influence the transduction process.

2. Fluctuation-Induced Instability: A Model for Natural Chemosensors

It was shown by Tomita and Tomita[6] that in a far from equilibrium system involving thermodynamically coupled degrees of freedom, limit cycle behavior can arise, induced by fluctuations from the steady state. They termed the situation resulting in a Hopf bifurcation as "ireversible circulation of fluctuations" characterizing the breakdown of detailed balance associated with birth of the limit cycle. Their result for chemically reacting systems involving at least two thermodynamically coupled degrees of freedom was deduced by considering the set of quantities constituting the system $\{X_i\}, (i = 1, 2, \dots n)$ as being characterized by Markov processes. The probability distribution $P(X, t)$ for the system in a given configuration then obeys the master equation:

$$
\begin{aligned}
\frac{\partial P(X, t)}{\partial t} &= \int dX' (P(X', t) W(X' \to X) \\
&\quad - P(X, t) W(X \to X')) \\
&= \int d(\Delta X)(P(X - \Delta X, t) W(X - \Delta X, \Delta X) \\
&\quad - P(X, t) W(X, \Delta X))
\end{aligned}
\tag{1}
$$

where $W(X, \Delta X)$ is the transition probability per unit time for a transition from X to the state $X + \Delta X$, denoted by X'. Scaling the variables as: $X = \epsilon^{-1}x$ where ϵ is the reciprocal of the system size parameter, they expressed the stochastic variable $x(t)$ as a departure from the deterministic one $y(t)$:

$$
x = y(t) + \epsilon^{1/2}\xi
\tag{2}
$$

with a distribution function in ξ as:

$$
p(\xi, t) = \Omega^{d/2}\psi((y(t) + \epsilon^{1/2}\xi), \ t)
\tag{3}
$$

Using a Kramers–Moyal expansion of the right hand side of Eq. (1), they showed that to lowest order the master equation reduces to the linear Fokker–Planck equation:

$$
\frac{\partial}{\partial t}p(\xi, t) = -\frac{\partial}{\partial \xi}G(\xi, t)
\tag{4}
$$

where G is the probability flux G given by

$$G(\xi,t) = K(y)\xi p(\xi,t) - \frac{1}{2}\frac{\partial}{\partial\xi}D(y)p(\xi,t) \qquad (5)$$

K and $D(t) = c_2(y(t))$ being the regression and diffusion matrices respectively, where $c_2(x) = \int d(\Delta X)(\Delta X)^2 w(x,\Delta X)$ and $w(X,\delta X) = \epsilon W(X,\delta X)$.

Tomita *et al.*[6] showed that the antisymmetric part G_a of G, defined through $G_a = \alpha \cdot \frac{\partial p(\xi,t)}{\partial\xi}$, (where $\alpha = \frac{1}{2}(K^T\sigma - K\sigma)$, with variance $\sigma = \int \xi_i\xi_j P(\xi,t)d\xi$), is rotational in nature and always non-vanishing in a non-equilibrium system. The antisymmetric matrix α quantifies the instantaneous directed circulation of the fluctuation and is a measure of the deviation of the system from detailed balance. The eigenvalues of the matrix K determine the nature of the system stability near the fixed points of the flow. When the eigenvalues are complex conjugate with vanishing real parts, the system undergoes a Hopf bifurcation which is fluctuation-induced.

This result of Tomita *et al.* confirms that fluctuations would play a dominant role in determining the outcome of far from equilibrium situations, in particular, those of biochemical reactions and consequently, would vastly influence the efficacy of the sensory cells in responding to various stimuli.

We describe chemosensory processes using nonlinear excitable elements (system variables could be, say, concentrations of the enzymes, ligands and receptors involved), each admitting the possibility of fluctuation-induced limit cycle behavior arising from a Hopf instability.[7] Upon incidence of the stimulus, ligand-receptor binding takes place, setting in motion chemical signalling reactions and cascades. These reactions are controlled by regulatory feedback mechanisms which come into play due to changes in concentration of the initial reactants. It is known that the signal transduction mechanisms for most (naturally occurring) chemosensors are regulated by G proteins and calcium ions.

We consider the simplest possible situation involving concentrations of just two species u and v, and one control parameter C (for instance C could be concentration of calcium ions or concentration of G proteins) which is a function of u and v.[7] The system would then be described by the set of equations:

$$\dot{u} = f(u,v,C), \quad \dot{v} = g(u,v,C), \quad \dot{C} = h(C,u,v) \qquad (6)$$

Following Tomita *et al.*'s work, we assume that the signal detection reaction network has at least one limit cycle from a Hopf bifurcation which is fluctuation-induced. This assumption is useful since, as mentioned above, it is well known that a nonlinear system operating near a Hopf bifurcation, shows an amplified response — a requirement for a good signal detector. Moreover, this also helps to account for spontaneous oscillatory activity experimentally observed[7] in sensory cells.

The canonical normal form equation for such a Hopf oscillator is

$$\frac{dz}{dt} = A(\omega, C)z - B(\omega, C)|z|^2 z + O(|z|^4 z) \tag{7}$$

where $z \in \mathbf{C}$, A, B are complex coefficients which depend upon the control parameter C and the frequency of the limit cycle ω.

Equation (6(iii)) describes the self-regulatory mechanism of the control parameter C which continually tunes the system to operate in the close vicinity of the Hopf bifurcation.

Any attempt to describe detection of a chemical sense within a mathematical model is immediately confronted with a problem: how could one quantify attributes such as odour or taste carried by a stimulus?

It was shown in Ref. 7 that for such spectral senses where receptors detect the chemical stimuli by their molecular vibrations, the applied stimulus can be taken to be a pulse lying between two very closely spaced frequencies ω_1 and ω_2, with very small bandwidth $\Delta\omega = \omega_2 - \omega_1$, and average frequency $\omega_{av} = \frac{\omega_1 + \omega_2}{2}$.

In the context of the spectral theory of olfaction (Turin[8]), we took the applied pulsatile stimulus $F_{ext} = \psi_0(t) \cos\omega_{av}t$, (where $\psi_0(t) = \frac{F_0}{2\pi^{1/2}\alpha}e^{-(t-\frac{\tau}{2})^2/\alpha^2}$, with $\tau \gg \alpha$, α being small) to have frequency equal to the vibrational frequency of the odorant molecule which binds to an olfactory receptor. The external stimulus activates the signal detection and transduction pathway. Our conjecture detailed in[7] is summarized in Fig. (1).

The perturbed Hopf oscillator is then described by:

$$\frac{dz}{dt} = A(\omega, C)z - B(\omega, C)|z|^2 z + \psi_0(t)\cos\omega_{av}t + O(|z|^4 z) \tag{8}$$

An additional component of fluctuations due to interaction with the external environment enters the system, and this could be represented by a weak additive Gaussian white noise $\zeta(t)$.[7] $\zeta(t)$ need not arise from thermal

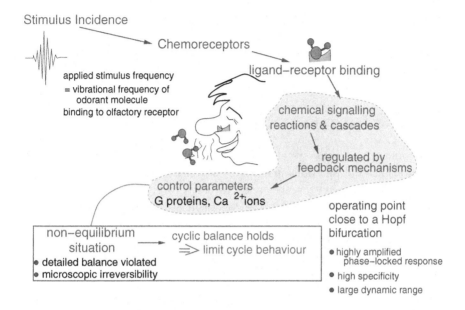

Fig. 1.   Detection of an odorant molecule by an olfactory receptor described in detail in Ref. 7 in a generic way for a chemosensor.

noise alone, but includes also other random influences such as fluctuations in vibrational modes of neighboring molecules, fluctuations in enzymatic cascades of transduction, etc.

The perturbed Hopf oscillator's response to a stimulus was calculated in Ref. 7 and it was shown that it is phase-locked to the center frequency of the external stimulus, and very hugely amplified in the presence of noise.

Also the spectral density $S(\psi)$ of this feedback-controlled system was calculated in the absence of stimulus[7] and it was shown to follow the power law $S(\psi) = \psi^{-\gamma}$ where γ lies between 1.2 and 1.4. Since noise encountered in a wide range of biological processes is near pink ($\gamma = 1$), we think that this result is of particular interest.

Behavioral features of chemosensory receptors (or other complex living structures) that are seen experimentally, may thus be reproduced usefully with appropriately designed minimal mathematical models. All the observed features — highly amplified phase-locked response to stimuli in a frequency-selective manner, partial and perfect adaptation to the stimulus, spontaneous activity whose spectral power density indicates pink noise usually associated with biological systems — emerge from a single generic model and exploit the premise that the sensory cell operates close to a dynamical instability.

3. Type-I and Type-II Neuronal Oscillations

An organism's nervous system and brain are able to perform myriads of computations and functions in spatially segregated regions simultaneously. All this data is pieced together and processed to give rise to features and streamlined information which the organism is able to perceive and act upon. How these impressive feats of information collection, processing and decision making occur, have always perplexed researchers over the years. Moreover, how are different neuronal interconnections and firing patterns related to different kinds of behavior?

Synchronization of neuronal activity has been thought to play an important role in feature extraction, recognition and perception. Differing kinds of synchronization phenomena exist — phase synchronization, frequency synchronization, and complete synchronization, the latter implying synchronization in phases, frequencies and amplitudes. It is possible that studies of synchronization mechanisms in the activities of coupled neurons, or changes in patterns of firing or bursting behavior could shed some light upon how information is selectively extracted and perceived.

Instabilities govern the firing patterns of single and coupled neurons. It has come to be recognized that information processing by neurons depends not just upon their electrophysiological properties, but on their dynamical properties as well. A classification of neurons on the basis of excitability patterns of the axon was done by Hodgkin and Huxley.[9] They termed as type-I excitability, neuronal activity involving transition from quiescent state to periodic spiking through a dynamical instability known as saddle-node bifurcation on an invariant circle. Such oscillators, also called integrators in the literature[10,11] can have an arbitrarily small emerging frequency depending upon the strength of the applied current. In type-II excitable neurons, also called resonators,[10,11] the transition from quiescent to a periodically spiking state is through a Hopf bifurcation, the action potentials being generated in a frequency band that is relatively insensitive to changes in the applied current.

Studies show[4,5,10-12,15] that the presence of noise in the system can induce synchronous activity in neurons. In our studies of complete synchronization of coupled type-I neurons, we found[4,5] that the presence of weak noise changed the operating point of the system and induced an instability for an existing set of control parameters. This produced a delay in the decay time of the synaptic gating variables, eventually producing complete synchronization. Therefore changes in the dynamical behavior of single or coupled sensory systems are brought about by the presence of instabilities. These changes dictate the future firing patterns or synchronous behavior of neurons and therefore would be expected to have a profound influence on the organism's perception and cognition.

4. Conclusions

The importance of understanding the dynamical properties of sensory cells and structures is emphasized here. Understanding the instabilities underlying the dynamical operation of sensory cells could well prove to be a useful, path breaking method for gaining insights into the unfathomed realms of cognition and consciousness. Techniques of dynamical systems theory may be fruitfully employed to understand how the presence of an instability could help in stimulus detection, feature extraction, etc.

References

1. V.M. Eguiluz, M. Ospeck, Y. Choe, A.J. Hudspeth, M.O. Magnasco, *Phys. Rev. Lett.* **84**, 5232 (2000).
2. R. Stoop, R. and A. Kern, *Phy. Rev. Lett.* **93**, 268103 (2004).
3. M. Ospeck, V.M. Eguiluz, and M.O. Magnasco, *Biophys. J.* **80**, 2597 (2001).
4. L. Turin, *Chem. Senses* **21**, 773 (1996); L. Turin, *J. Theor. Biol.* **216**, 367 (2002); G.M. Dyson, *Chem. Ind.* **57**, 647 (1938); R.H. Wright, *J. Theor. Biol.* **64**, 473 (1977).
5. J.C. Brookes, F. Hartoutsiou, A.P. Horsfield, and A.M. Stoneham, *Phys. Rev. Lett.* **98**, 038101 (2007).
6. K. Tomita and H. Tomita *Prog. Theor. Phys.* **51**, 1731 (1974); K. Tomita, T. Ohta, and H. Tomita, *Prog. Theor. Phys.* **52**, 1744 (1974); K. Tomita, T. Todani, and H. Kidachi, *Physica A* **84**, 350 (1976).
7. J. Balakrishnan and B. Ashok, *J. Theor. Biol.* **265**, 126, (2010).

8. N. Suzuki, *Comp. Biochem. Physiol. A Mol. Integr. Physiol.* **61**, 461 (1978); D. Baylor, G. Matthews and K-W. Yau, *J.Physiol.(Lond)*, **309**, 591 (1980); P. Duchamp-Viret, L. Kostal, M. Chaput, P. Lansky, and J-P. Rospars, *J.Neurobiol.*, **65**, 97 (2005).

9. A.F. Huxley and A.L. Hodgkin, *J. Physiol. (London)* **117**, 500 (1952).

10. E. Izhikevich, *IEEE Trans. Neural Networks* **10**, 499 (1999).

11. B. Ermentrout, *Neural Comput.* **8**, 979 (1996).

12. A. Pikovsky, M. Rosenblum and J. Kurths, *Synchronization: A Universal Concept in Nonlinear Sciences*, Cambridge University Press, Cambridge (2001); M. Lakshmanan and S. Rajasekar, *Integrability, Chaos and Patterns*, Springer, (2002).

13. N. Malik, B. Ashok and J. Balakrishnan, *Pramana — J. of Phys.* **74**, 189 (2010).

14. N. Malik, B. Ashok and J. Balakrishnan, *Eur. Phys. J. B* **74**, 177 (2010).

15. C. Börgers and N. Kopell, *Neural Comput.* **15**, 509 (2003); *Neural Comput.* **17**, 557 (2005).

Chapter 6

Active Cellular Mechanics and Information Processing in the Living Cell

M. Rao

Raman Research Institute, CV Raman Avenue, Bangalore 560080, India and National Centre for Biological Sciences (TIFR), GKVK Campus, Bellary Road, Bangalore 560065, India

rao.madan@gmail.com

I will present our recent work on the organization of signaling molecules on the surface of living cells. Using novel experimental and theoretical approaches we have found that many cell surface receptors are organized as dynamic clusters driven by active currents and stresses generated by the cortical cytoskeleton adjoining the cell surface. We have shown that this organization is optimal for both information processing and computation. In connecting active mechanics in the cell with information processing and computation, we bring together two of the seminal works of Alan Turing.

1. Introduction

I will present recent work on the active organization of signaling molecules on the surface of living cells, its mechanical basis and their consequences to information processing and cellular computation.

To start with, let us see why there might be a deep connection between the molecular organization, active cellular mechanics and information processing and computation in the cell. The *cell*, the basic unit of life, is a collection of specific biomolecules with a defined n spatiotemporal organization. This organization is self-regulated at different scales, utilizing energy available in a variety of forms. Since all that is there in the cell are molecules, it is apparent that "information" is encoded in the state of the biomolecules and their organization. This would immediately suggest that the organization, transport and chemical transformation of biomolecules, which is driven by the active (energy-dissipating) mechanics of the cellular milieu, in turn, drives the storage, flow and processing of information. It is in this sense that the cell translates chemical and physical processes into the management of information.[1]

In connecting active mechanics in the cell and tissues with information processing and computation, we bring together two seminal works of Alan Turing — his first, the 1937 paper on the *theory of computation*, and his last, the 1952 paper on the *chemical basis of morphogenesis*.[2]

2. Cell Surface as an Active Medium

The localization, transport and chemical modification of molecules take place in a cellular medium which is driven by thermal and active stresses, arising primarily, though not solely, from motor-cytoskeletal filament complexes, such as myosin-actin and kinesin-microtubule, and involve the continuous consumption of chemical energy. Every part of the cell is subject to such active processes — the cytoplasm, the nucleus and the cell surface.

In collaboration with biologist Satyajit Mayor's group at NCBS, we have engaged in an experimental and theoretical program to understand the physico-chemical principles governing molecular organization and dynamics on the cell surface at a variety of scales. Ourselves[4] and others[5] have found that many cell surface signaling receptors are organized as dynamic nanoclusters driven by the active mechanics of the cortical actin-myosin adjoining the cell surface, and have developed a theoretical framework to understand the dynamics of local composition and shape of the active composite cell surface. This involves a paradigm shift, from thinking of the cell surface as a closed system at thermodynamic equilibrium to an *open* system driven out-of-equilibrium by the consumption of chemical energy available to the cell. We have proposed new organizing principles underlying the spatiotemporal regulation of local composition and shape — the *Active Composite Model* of the cell surface.

The active composite model[7] assumes that the cell membrane adjoins a cortical actin which comprises of a static, crosslinked meshwork *in addition* to a pool of highly dynamic and active actin filaments (driven by actomyosin contractility and treadmilling). The steady states of the dynamic component of the actin cortex consist of localized contractile elements (asters) that dynamically remodel. The effect of the active cortex on cell membrane composition is best studied by classifying them as *inert, passive* and *active*, based on their engagement with this active cytoskeleton. While inert molecules may at best be advected by hydrodynamic flows generated by active stresses, they do not directly couple to actin. On the other hand, passive molecules, by virtue of their coupling to the dynamical actin filaments, form transient nanoclusters by being focussed into the remodeling asters. The active composite model has been quite successful in explaining and predicting the many characteristic features of molecular organization of passive cell surface molecules that engage with cortical actin.[7] Finally, active molecules, such as Integrin and T-cell signaling receptors, are both driven by and, in turn remodel, cortical actomyosin. Active molecules can nucleate, destroy or induce coalescence and flows of actin asters. In the cellular context, the identity of molecules as inert, passive and active can be altered by conformational switching.

3. Cell Surface as a Distributed System for Information Processing and Computation

The underlying mechanics of the active composite model, naturally leads to a classification of molecules as inert, passive and active. In addition, the model brings forth an *emergent particle*, namely localized focusing platforms (or asters), which can draw in other molecules and transiently focus them within its core. This sets the stage for asking how chemical transformation of molecules or chemical kinetics might be affected by this underlying active mechanics.

We find that the transient focusing of passive molecules into nanoclusters by active contractility and remodeling of the cytoskeleton, leads to a dramatic increase in the reaction efficiency and output levels.[9] The dynamic cytoskeletal elements that drive focusing behave as *quasi-enzymes* catalyzing cell surface chemical reactions. We have proposed that such cytoskeletal driven clustering of proteins could be a cellular mechanism to spatiotemporally regulate and amplify local chemical reaction rates in cell surface signaling.

Chemical transformation of biomolecules is the basis of cellular information processing. It is natural to expect that the cell has evolved to optimize specific aspects of this information processing. We have recently shown that these active actin-dependent mechanisms of localization, formation and breakup of nanoclusters of cell surface molecules have deep implications for information processing in the cell. For instance, we have studied the optimum strategies for a collection of protein sensors on the cell surface to cooperatively estimate an external signal, such as a ligand field, that is a

function of space and time. We demonstrate a *phase transition* in the space
of strategies, as a function of sensor density and efficiency — at high sensor
densities the optimal solution is to arrange the sensors on a regular lattice
grid (e.g., chemotactic receptors in bacteria), while at low densities the optimal solution is provided by the actively driven focusing and remodeling of
transient clusters (e.g., cell surface receptor clustering in eukaryotic cells[7]).
To realize this dynamic strategy, the cell surface needed to be relieved of
the constraints imposed by the rigid scaffold, and to be more regulatable.
The strategy change needed the innovation of motor proteins and dynamic
actin filaments, a regulated *actomyosin* machinery, fueled by chemical energy in the form of ATP. In this case, the cell machinery has evolved to
optimize the faithful reading of an incoming function.

How does the information received at the cell surface translate to down-
stream signaling? This involves reading and interpreting information from
the outside (input) and transmitting it within the cell (output), before re-
setting itself for another round of processing; in other words, *computation*.
The 'computational hardware' comprises of specialized protein receptors,
whose dynamic clustering in transient signaling platforms (asters) and ac-
tivation status is spatiotemporally regulated by the same ATP-dependent
active mechanics of the active composite cell surface described above. These
protein receptors can either be drawn into pre-existing signaling platforms
(passive molecules) or can induce the formation of signaling platforms (active molecules). This forms the basis for an actively regulatable Turing
computation, enabling a variety of external inputs to be transduced into
signaling outputs, through the construction of logic gates such as AND, OR
and NAND. We construct measures of computational efficiency and show
that these are enhanced by the active driving. We propose that the underlying mechanics of the active composite cell surface allows it to function
as a reconfigurable programming platform. Our study has relevance in a
variety of contexts involving cellular computation of external inputs, such
as in integrin and T-cell signaling.

4. Conclusion

It appears from these studies that active cellular mechanics may provide
the underlying mechanism to translate physical processes to the regulation
and management of information.[1] It is in this sense that one may talk about
the mechanics of cellular computation and information processing. In the
context of the cell surface, we argue that the physical principles underlying

the active composite model of the cell surface provides the natural language for discussing the mechanics of computation and information processing at the cell surface.

I have highlighted our recent work on the interplay between active cellular mechanics and information processing and computation via cellular biochemistry. The implication is that the cellular machinery would have evolved to arrive at optimal solutions to information processing problems posed within a given niche or environment. We are currently exploring strategies on how to test these ideas experimentally. It is possible that extensions of these active mechanical principles might also be involved in a collection of neuronal cells, specialized to process information of a particular kind.

5. Acknowledgements

I thank my numerous collaborators and students, especially S. Mayor, G. Iyengar, K. Husain and S. Krishna, for help in sharpening some of the ideas presented here.

References

1. P. Nurse, The great ideas of biology, *video lecture*, http://royalsociety.org/events/2010/great-ideas-biology/.
2. A.M. Turing, On Computable Numbers, with an Application to the Entscheidungsproblem, *Proc. London Math. Soc.* **2**:230-265 (1937); The Chemical Basis of Morphogenesis, *Phil. Trans. Roy. Soc.* B **237**:37-64 (1952).
3. M.C. Marchetti, et al., Hydrodynamics of Soft Active Matter, *Rev. Mod. Phys.* **85**:1143-1189 (2013).
4. S. Mayor and M. Rao, Rafts: Scale-dependent, active lipid organization at the cell surface, *Traffic* **5**:231-240 (2004).
5. J.F. Hancock, Lipid rafts: contentious only from simplistic standpoints, *Nat. Rev. Mol. Cell Bio.* **7**:456-462 (2006).
6. A. Kusumi et al., Dynamic organizing principles of the plasma membrane that regulate signal transduction: commemorating the fortieth anniversary of Singer and Nicolson's fluid-mosaic model, *Ann. Rev. Cell and Dev. Biol.* **28**:215-250 (2012).
7. K. Gowrishankar et al., Active remodeling of cortical actin regulates spatiotemporal organization of cell surface molecules, *Cell* **149**:1353-1367 (2012).
8. G. Iyengar and M. Rao, Cellular Solution to an Information Processing Problem, *2012 IEEE Information Theory Workshop, Lausanne,* Sept. 3–7 (2012); and Active clustering of cell surface receptors: solution to a signal processing optimization problem, *manuscript in preparation.*

9. A. Chaudhuri et al., Spatiotemporal regulation of chemical reactions by active cytoskeletal remodeling, *Proc. Nat. Acad. Sc.* **108**:14825-14830 (2011).

10. K. Husain, S. Krishna, S. Mayor and M. Rao, Cell surface as an active computing platform, *manuscript in preparation.*

Chapter 7

On the importance of length scales in determining the physics of biological systems

B. Ashok

Centre for Complex Systems & Soft Matter Physics, International Institute of Information Technology, Bangalore (I.I.I.T.-B), 26/C, Electronics City, Hosur Road, Bangalore 560 100, India bashok@iiitb.ac.in, bashok1@gmail.com

We discuss the effect and importance of electrostatic and other interactions in real life systems. We note that the consideration of appropriate length scales is crucial in discussing dynamical effects and see how this influences information transfer in diverse systems, including biological systems.

1. Introduction

Information transfer occurs in various systems in nature on diverse length scales in various forms. What effects get manifested and how crucially depends on the interactions involved and the length and time scales that are being investigated. In the following paragraphs, we shall briefly discuss some of the phenomena seen in nature, especially in the context of biological systems, and mention the theoretical models used to explain them.

In nature, all the fluids that are present in cells are essentially "Coulomb soups" — polyelectrolytic and electrolytic solutions of varying concentrations.

Even within the cell, apart from having complex structures like Golgi bodies and microtubules, the nuclear envelope has several nuclear pore complexes (of the order of about 20 to 100 nanometres) formed by the self-assembly of various proteins. It is through such structures that mRNA and mRNP complexes which are typically of size 50nm thread themselves. These phenomena occur at large time and length scales as compared to atomistic scales. The characteristic length scale of a system plays a crucial part in the physics of the system.

Macromolecules or polymers are ubiquitous in nature: everything from RNA to DNA to microtubules to larger systems comprise of long chain molecules. Hence understanding phenomena such as DNA translocation at the cellular level calls for a treatment of behavior of such large chains that falls under the purview of polymer physics. A very good treatment of the physics behind such phenomena can be found in, for example, a recent work on polymer translocation and polyelectrolyte dynamics,[1] and references therein.

In nature, the electrostatic interactions between molecules, between cells, and within various parts of living organelles play an important role in the functioning of living organisms. The DNA chain, for example, is a polyelectrolyte since its backbone is negatively charged. When DNA/RNA is packed within a viral capsid, the chain takes a coiled, ellipsoidal configuration. That electrostatics plays a crucial part in causing such a configuration has been proven through Brownian dynamics simulations. The dynamics of DNA chains is described through both consideration of the electrostatic interactions involved as well as the extent of flexibility of the chain, as well as, for example, the hydrodynamic and other interactions present in the system.

2. Macromolecules and Polyelectrolyte Solutions

The behaviour and dynamics of macromolecular solutions is determined by various factors. The rigidity of the chains are important — whether the polymer chains are flexible, semiflexible or stiff chains. Another factor that determines system behavior is which concentration regime we are dealing with — whether the solutions are dilute, semi-dilute or concentrated. Yet another important factor that drastically influences dynamical and configurational properties is whether the chains are neutral or charged, i.e., whether the macromolecules are uncharged polymers or whether they are polyelectrolytes.[2-5]

In the dilute case, hydrodynamic interaction predominates. The excluded volume interaction, as also the electrostatic interaction in the case of polyelectrolytes, are additional features. When a polymer or polyelectrolyte solution is in the semi-dilute regime, its dynamics may vary as several of the interactions affecting the chains may be screened off to different extents. Intrachain entanglements also may be present in concentrated polymer solutions. In the references mentioned above and in the literature cited therein, varied approaches have been taken to tackling the problem of a theory for polyelectrolyte solutions.

Let us consider the case of dynamic properties of polymer solutions such as viscosity. The shear and bulk viscosity of a material are measures of the translational and rotational friction coefficients for that substance. Knowledge of the viscous properties of a fluid is clearly very important, since this directly impinges on the motion of any object through the fluid.

What is the quantity knowing whose behavior we can get an immediate handle on the viscosity of a macromolecular solution? This entity, it turns out, is the persistence length of the polymer or polyelectrolyte. The persistence length is that distance of the polymer chain over which tangents drawn along the arc distance retain their direction and correlation. In polymer physics, the length scale often referred to is the Kuhn length l, which is twice the persistence length. Each Kuhn segment can consist of several monomer units. The expression for the translational friction coefficient ζ for a polyelectrolyte with molecular weight N (i.e., with N segments of repeating units in the polyelectrolyte chain) is given by

$$\zeta = 3\eta_0 \sqrt{Nll_1}/(8\sqrt{2}), \tag{1}$$

with ζ in turn being intimately connected to the shear viscosity. Here, l_1 is what is known as the renormalized Kuhn length — a measure of the effective Kuhn length of the polymer chain in the presence of electrostatic and other interactions, while l is the bare Kuhn length (for the ideal, Gaussian chain — Gaussian since the conformational distribution function obey a Gaussian distribution). η_0 is the shear viscosity of the solution in the absence of the chain. Thus, $l_1 \geq l$. The radius of gyration R_g, defined through

$$R_g^2 = Nll_1, \tag{2}$$

gives us a measure of the length of the polymer chain. Figure 1 gives a sketch showing the various basic lengths used in describing a polymer chain — the persistence length, the Kuhn length and the radius of gyration.

When we are considering several charged particles and chains, the net effect due to Coulomb interactions get screened and is replaced by a screened electrostatic potential of the form $V \sim \exp(-\kappa r)/r$, with the

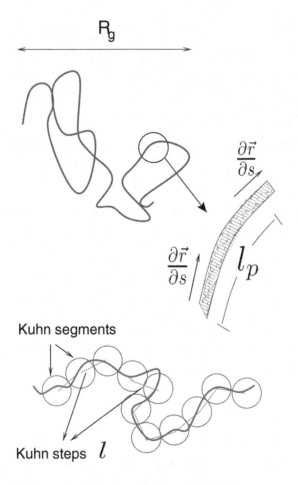

Fig. 1. Schematic showing a polymer chain, with radius of gyration & persistence length shown, as also a Kuhn segment model of a chain.

Debye screening length κ^{-1} giving the length scale to which the electrostatic interaction is effectively felt in the presence of a large number of charged ions. κ^2 is proportional to the counterion and salt concentrations. One routine comparative measure that is often used to quantify the electrostatic interaction in a system is the Bjerrum length l_B given by

$$l_B \equiv e^2/k_B T \epsilon, \tag{3}$$

where ϵ is the electric permittivity for the fluid. This is the distance where the potential energy of the electrostatic Coulomb interaction equals the thermal energy $k_B T$.

On including the excluded volume interaction (whose strength is denoted by some parameter w) between the chains, i.e., we are accounting for the fact that chains cannot interpenetrate each other and there will also be an effective interaction between the chains, one more length scale, the Edwards screening length ξ_e, comes into play. This is the distance to which the excluded volume interaction is effective, after which for $x > \xi_e$, the influence of this interaction falls off exponentially. This ξ_e is proportional to $c_p^{-3/4}$ in the semidilute regime where c_p is the polymer concentration, and also depends upon κ.

Another length scale that plays an important role in the system's dynamic behaviour is the hydrodynamic screening length

$$\xi = (2/\pi)(c_p l l_1)^{-1}, \tag{4}$$

where c_p is the polyelectrolyte number density. This comes into play when fluid interactions and hydrodynamics become important in the system, and gives the spatial extent to which hydrodynamic interactions can be felt.

For $\kappa R_g > 1$,

$$\xi = (2/\sqrt{\pi})(1/3^{3/4})(w + (4\pi l_b/\kappa^2))^{-1/4}c_p^{-3/4}l^{-1/2}. \tag{5}$$

However, when $\kappa R_g < 1$, ξ has the form

$$\xi = (8/\sqrt{3}\pi)(\pi/6\sqrt{2})^{2/3}(l/4\pi l_b)^{1/6}c_p^{-1/2}l^{-1/2}. \tag{6}$$

Details about the treatment of screened interactions, screening lengths, etc., may be found in, for example, Refs. 2–5. All in all, there is a very complex interplay between the various length scales involved, far too involved at times to handily sort out individually and identify as the prime cause for a given effect.

Thus on comparing polyelectrolyte solution behavior to that of uncharged polymer solutions, various differences are at once evident in experiments.

The reduced viscosity, η_{red}, of a solution is defined as

$$\eta_{red} \equiv (\eta - \eta_0)/(\eta_0 c_p), \tag{7}$$

where η is the solution viscosity, η_0 is the solvent viscosity and c_p is the polyelectrolyte concentration. When the reduced viscosity of polyelectrolyte solutions is plotted as a function of polymer concentration, a peak appears at low concentrations, which is absent in the case of uncharged polymer solutions. Increasing the added-salt concentration, i.e., increasing counterion concentrations, caused a suppression in the reduced viscosity peaks. Indeed, until some years back, there was no satisfactory theory explaining the cause of the experimentally observed facts.

A field-theoretic theory that makes use of an effective medium approximation within a multiple-scattering approach finally resolved this longstanding question in the polyelectrolyte solutions literature.[5] This theoretical approach was robust in that it was also successfully able to theoretically duplicate experimental observations seen for uncharged polymers, while also predicting the effect of various parameters such as the degree of ionization of the solution on the viscosity. While we shall not dwelve into the details here, suffice it to say that the principal cause for the markedly different behavior for polyelectrolyte solutions turns out to be the electrostatic interaction between charged chains, even while competing effects between hydrodynamic, excluded volume and electrostatic interactions — both inter- as well as intra-chain — as mentioned above, contribute to the final effect seen. Thus while at very low polyelectrolyte concentrations, the uncharged and charged polymer solutions behave the same, at the crossover regime between dilute and semidilute solutions, there is a drastic change in the polyelectrolyte reduced viscosity curves with a peak appearing. Figure 2 shows a schematic sketch of the phenomena seen.

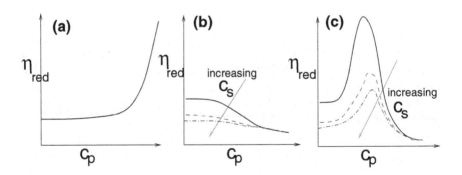

Fig. 2. Reduced viscosity η_{red} plots as a function of polymer concentration c_p for uncharged polymer (a) and polyelectrolytes (b) without accounting for interchain electrostatic interaction and (c) after inclusion of interchain electrostatic interaction.

A question that naturally comes to mind is this: in what way does the presence of a polyelectrolyte/electrolyte solution in biological structures influence routine behavior as opposed to having just a typical, Newtonian fluid? Does this facilitate daily activities that are essential for the existence of life in the organism? It has already been observed that certain phenomena at the cellular level, such as threading of RNA through the nuclear pore complex, as also the encapsulation and release of DNA from a viral capsid, and other similar phenomena involving polymer translocation, involves the effects of electrostatics. It would seem that there is an ideal range of parameter values within which motion is facilitated for long chain charged molecules, there being a large, robust barrier in the shape of an enhanced viscosity to concentration ratio at higher polyelectrolyte concentrations that discourages motility in a cross-over regime. This perhaps helps in the regulation of transfer phenomena in living cells and structures so that it is only in optimum "Coulomb soup" conditions is information transfer or suitable physical motion of the molecules facilitated.

From the discussion above, we could therefore expect that the viscosity of a fluid or liquid medium at small scales may not necessarily have a uniform value and would, instead, depend on the length scales being probed. This has, in fact, been shown to be true. In Ref. 6, the viscosity of the cytoplasm in mice muscle cells and human cancer cells are reported. They measure the correlation length in the fluid as well as the hydrodynamic radius and find that with probes larger than the hydrodynamic radius, the fluid viscosity is the macroscopic viscosity. At length scales below the hydrodynamic radius but larger than the correlation length, the viscosity increases greatly with probe size.

3. Motion of Micro-Organisms in Fluids

One of the parameters that helps quantify in viscous media is the Reynolds number, $\mathcal{R}$. The Reynolds number is the ratio of the inertial forces acting on a body to the viscous forces. Hence, if the body being described has dimensions of the order of r, and is moving through a fluid with velocity v, then

$$\mathcal{R} = rv\rho/\eta, \tag{8}$$

with ρ being the fluid density and η its viscosity. For water, $\eta \approx 10^{-3}$ Pa s, and $\rho \approx 10^3$ kg/m^3. For organisms of dimensions $\sim 1\mu$m, moving at

$v \approx 10 \, \mu\text{m/s}$, $\mathcal{R} \approx 1 \times 10^{-5}$. At such low Reynolds numbers, the inertial contribution to motion becomes insignificant. Thus any motion of an object at a given time becomes fully determined by the forces at that instant, independent of its past history. Movement or swimming in a fluid can be achieved by some cyclic deformation of the organism. Motion in fluids is described by the Navier–Stokes equation

$$\nabla p + \eta \nabla^2 \boldsymbol{v} = \rho \frac{\partial \boldsymbol{v}}{\partial t} + \rho (\boldsymbol{v} \cdot \nabla) \boldsymbol{v}, \qquad (9)$$

with p being the pressure, v the velocity, and ρ the fluid density.

Now since the system is at low Reynolds numbers, we can neglect the inertial terms in the Navier–Stokes equation describing motion in a fluid. We are therefore just left with the equation $\nabla p = \eta \nabla^2 \boldsymbol{v}$. Clearly, if any cyclic deformation occurs that involves just reversing the sequence of steps in time, nothing will happen as time plays no role in determining the motion. If an organism were to have stiff arms that it used exactly as oars to try and row through the fluid at low $\mathcal{R}$, it would be in trouble. If, however, these arms were *flexible*, the rowing motion would vary as in the return stroke the direction of bending of the arm would be opposite to that in the forward stroke, and locomotion would be possible. The reason why several micro-organisms have cilia and flagella now becomes clear — for without these aids, movement becomes impossible.

The above brief discussion on the movement of micro-organisms has been adapted from the eminently readable article of Purcell[7] on motion at low Reynolds numbers, to which the reader is referred to for details. It would be appropriate to note at the conclusion of this section that the question of motion of an organism in a fluid was first considered by Taylor in his classic paper.[8]

4. Microtubules and Plant Growth

Microtubules are important constituents of various living organisms. They consist of subunits of the tubulin protein (heterodimers of α- and β-tubulin) arranged helically around a hollow core in long protofilaments, such that these are polar structures, 10 to 25 nm in diameter, and about 10μm long.

In plants, cell development can be influenced greatly by cortical microtubules which form a part of the cytoskeleton in plant cells during interphase, the stage preceding and following cell mitosis. The direction and orientation of microfibrils influences the direction of cell expansion. Since microtubules provide an oriented scaffolding under the cytoplasm that is parallelled by the growing cellulose, the microtubule orientation essentially determines the direction of cell expansion. Moreover, alignment of cell division is determined by the actual position of the cortical microtubules. Microtubules are also the backbone of the transport system by which vesicles (phospholipid-walled micron sized containers) and macromolecules traverse the cellular system.[9,10]

During a plant's growth, cortical microtubules reorient themselves in response to various events and environmental factors. One such factor is the effect of electric fields on the plant. It has been shown experimentally that electric fields affect microtubule alignment.

Further details on these can be found in, for example,[9] and the literature cited there.

This can be understood in part by an analogy with the effect of electric fields on confined polymers. In the case of diblock copolymers between two confining plates, application of an external electric field causes a realignment of the morphology to a minimum energy state. The Kuhn lengths of the copolymer chains determine whether the possible lamellar layers are commensurate with the distance of confinement, in the case of lamellar morphology. For cylindrical morphology, we get a limiting value for the possible ratio of permitted Kuhn lengths of the copolymer constituents for a real electric field to cause a change in orientation. If l_A and l_B denote the Kuhn lengths for the two components A and B, respectively, constituting the diblock, and f_A and f_B their respective compositions (i.e., $f_B = 1 - f_A$), then the ratio $\gamma \equiv (l_A/l_B)^2$, necessarily has to obey the constraint

$$\gamma \geq 8(1 - f_b)/f_B. \tag{10}$$

The above expression, of course, has been obtained for a simplified, ideal system in the strong segregation limit and has been discussed in detail in,[11] where expressions for the critical electric field required to cause a change in orientation of the block copolymer have been obtained. The other results of the theory require that there is a mismatch in dielectric constants of the two constituents of the diblock copolymer and that there is also a mismatch in the interfacial energy values between each of the components and the confining surface.

This equation and the above-mentioned treatment will not of course hold exactly when we consider the re-orientation of microtubules in an electric field, which is a much more complex system, but it nonetheless gives us some useful insights. For example, we could expect that the persistence lengths of the α- and β- tubulin monomers would not be identical, and that we could expect different values for their dielectric constants. We could in fact be extremely bold and try and see what their relative persistence lengths could be if we took $f_B \approx 0.7$, which, in block copolymer morphology phase diagrams, would correspond to a block copolymer with cylindrical and gyroid morphology. This gives $\gamma \approx 3.43$, which implies $l_A/l_B = \sqrt{\gamma} \approx 1.85$.

Now let us compare these estimates we have of the properties of the tubulin monomers with what is known in the microtubule and tubulin literature. As has been reported in,[12] there is, indeed, a substantial difference in the dipole moments and dielectric properties of α- and β- tubulin. Molecular dynamics simulations reported in the literature[13] yield elastic constants for α-tubulin and β-tubulin to be, respectively, in the ranges $k_\alpha \approx 3.9$ to 15.6 N/m and $k_\beta \approx 3.3$ to 11.5 N/m. Their ratio gives an indirect indication of the ratio of their persistence lengths to be, at the maximum, 1.35. Clearly, therefore, our bold shot in the dark was quite bold and substantially overestimated the ratio (by about 37%)! But it is interesting to note that as far as rough estimates go, this is not all that bad, given that data in the published experimental literature on the mechanical properties of tubulin proteins seem to span a very large range of values.

In plants, the meniscus in the mesophyll cell walls have dimensions of the order of $r \sim 10 - 40\mu$m. Since surface tension σ at the air-water interface at room temperature is about 70 dynes/cm, the pressure difference across the meniscus works out to be $\Delta P \approx 2\sigma/r \approx 1.4 \times 10^7$dynes /cm^2. This permits the rise of water to great heights and to the extremes of tall trees.[14] Clearly, the size of the stomata on a microscopic scale becomes a limiting factor in the macroscopic scale, limiting the extent of growth of the plant on the whole!

5. Concluding Remarks

In the preceding paragraphs, we have barely scratched the surface when it comes to discussing the details of the physics of interactions and information transfer in biological systems, and yet it can be seen that it has a rather daunting and non-trivial spread over various specialized disciplines.

Our approach here is necessarily restricted to a few samples that can hardly claim to span the whole gamut of phenomena and biological systems that are found in life. We have seen in various examples above, how Nature's choice of correct length scales becomes crucial for the survival of organisms.

It is interesting that dependence on such length scales also manifest themselves more generically and in a broader sense in nature. For example, there are several scaling arguments that are used to discuss the effect of environment on the size of animals. There is the "3/4 power law", where the metabolic rate scales as the 3/4 power of the body mass of the animal and is seen to hold for many organisms.

Then there is Bergmann's Rule, that seeks to generalize an observation that animals in cooler regions are typically larger than those in warmer climes. The argument for this is that larger animals would have a smaller surface area -to-volume ratio, so that radiative heat loss from the body is minimized, allowing them to maintain warmer body temperatures in the cold; the reverse would hold true for smaller animals whose higher surface area to volume ratio would aid heat loss and enable them to stay cooler in hotter climates.

The effects of and limitations due to natural length scales and size for animals have also been discussed at length in the literature, as also the origin of allometric scaling laws (see, for example[15-17]). The take-home message of this exercise clearly is that even while modeling the generic behavior of a system, it pays (indeed, it is essential) to attend to some specificities of the system, such that the model remains valid at all length and time scales. Another important point is that the various disciplines in the natural sciences have now sufficiently evolved to such an extent that we are now in a position to model (even if with limited success) really complex biological systems.

There is a quotation of Locke's, which it would be pertinent to repeat here: "It is of great use to the sailor to know the length of his line, though he cannot with it fathom all the depths of the ocean". To rephrase it from the physicist's point of view, one could say that it is of great use to the physicist to know the characteristic length of the system he is modeling, even if he may be unable thereby to quantify all the physical phenomena the system displays!

References

1. M. Muthukumar, *Polymer Translocation*, CRC Press, Taylor and Francis Group, Boca Raton (2011).
2. M. Muthukumar, *J. Chem. Phys.* **105**, 5183 (1996).
3. M. Muthukumar, *J. Chem. Phys.* **107**, 2619 (1997).
4. M. Doi and S. F. Edwards, *The theory of polymer dynamics*, Clarendon Press, Oxford (1986).
5. B. Ashok and M. Muthukumar, *J. Phys. Chem. B* **113**, 5736 (2009).
6. T. Kalwarczyk, N. Ziebacz, A. Bielejewska, E. Zaboklicka, K. Koynov, J. Szymanski, A. Wilk, A. Patkowski, J. Gapinski, H-J. Butt, and R. Hoyst, *Nano Lett.* **11**, 2157 (2011).
7. E. M. Purcell, *Am. J. Phys.* **45**, 3 (1977).
8. G. I. Taylor, *Proc. Roy. Soc. Lond.* A **209**, 447 (1951).
9. J. M. Hush and R. L. Overall, *J. Microscopy* **181**, 129 (1996).
10. P. H. Raven, R. F. Evert and S. E. Eichhorn, *Biology of Plants*, 7th Ed., W.H. Freeman and Company Publishers, New York (2005).
11. B. Ashok, M. Muthukumar and T. P. Russell, *J. Chem. Phys.* **115**, 1559 (2001).
12. J. A. Tuszynski, E. J. Carpenter, J. T. Huzil, W. Malinski, T. Luchko and R. F. Luduena, *Int. J. Dev. Biol.* **50**, 341 (2006).
13. S. Enemark, M. A. Deriu, M. Soncini and A. Redaelli, *J. Biomech. Eng.* **130**. 041008-1 (2008).
14. R. H. Rand, *Ann. Rev. Fluid Mech.* **15**, 29 (1983).
15. D. S. Glazier, *Proc. R. Soc. B* **275**, 1405 (2008).
16. K. Schmidt-Nielsen, *Scaling: Why is Animal Size So Important?*, Cambridge University Press (1984).
17. G. B. West, J. H. Brown and B. J. Enquist, *Science* **276**, 122 (1997).

Chapter 8

q-deformations and the dynamics of the larch bud-moth population cycles

Sudharsana V. Iyengar[1] and J. Balakrishnan[2]

[1] School of Physics, University of Hyderabad, Central University P.O.,
Hyderabad — 500 046, India.
[2] Complex Systems Group, School of Natural Sciences & Engineering,
National Institute of Advanced Studies, Indian Institute of Science Campus,
Bangalore 560012, India
janaki05@gmail.com

The concept of q-deformation of numbers is applied here to improve and modify a tritrophic population dynamics model to understand defoliation of the coniferous larch trees due to outbreaks of the larch bud-moth insect population. The results are in qualitative agreement with observed behavior, with the larch needle lengths, bud-moth population and parasitoid populations all showing 9-period cycles which are mutually synchronized.

1. Introduction

Although Nature abounds with complex phenomena, historically, the effort has always been to seek simple solutions and models to explain observed behavior. The simple harmonic oscillator has been used extensively in the literature, for instance, to model various situations; however it fails to capture complex phenomena which are by definition, nonlinear. With nonlinearities are associated a host of rich dynamical structures and phenomena such as nonlinear oscillations, chaos, fractals, etc. q-deformations of numbers and functions have been objects of study since long[1–4] and have been found useful in the literature for describing fractal and multifractal sets.[5] The limit $q \to 1$ recovers the original number/function. The physical interpretation of the deformation parameter q, however is still unclear. In this work, a possible application of q-deformed numbers in describing population models is proposed.

Multi-annual cycles have always fascinated ecologists for years. Explaining the remarkably precise periodicity observed in ecological systems has remained a big challenge. Many cycles have been attributed to physical effects such as temperature variation, sunspot activity, etc., while others have been attributed to interactions such as predator-prey. Recently maternal effects and genetic factors have also been shown to play a vital role in generating these cycles.[6]

We discuss here the dynamics of a population cycle which has been well documented over many years and which has been of intense interest.[7,8] The larch bud-moth *Zeiraphera diniana* is an insect which has been held responsible for periodically destroying the foliage of larch conifers in the Alps and in the North American continent. The insect outbreaks which have been known to occur with an 8 to 9 year periodicity have been noted in Europe over several centuries. In response to frost, variability in the available nutrient content, and the wasteful feeding by the bud-moths, the lengths of the larch needles also show a nine-year periodicity. The bud-moth during the course of its cycle is invariably preyed upon by parasitoids which attack all stages of its life-cycle except the adult stage.[8] The parasitoid infestation and changes in the food quality (protein and raw fibre content of the needles) are both believed to strongly regulate bud-moth populations.

We consider here a discrete-time tritrophic model[9,10] well known in the literature, for the larch needle — bud-moth — parasitoid system. We modify the equations to enable further flexibility in the system and apply different deformations to the three state variables. It is shown that in certain parameter regimes, all three variables — bud-moth population, parasitoid population and needle length, show regular oscillatory behavior, repeating after nine periods, in qualitative agreement with observations. In certain other regimes, different periodicities are possible, while the model also shows regimes where the system is chaotic. Period doubling and period-halving are also possible in some other parameter regimes.

2. q-deformations

Statistical mechanics rests on certain assumptions, such as the ergodic hypothesis, the extensivity of entropy, etc. The Maxwell-Boltzmann distribution describing the velocity of molecules in a gas in thermodynamic equilibrium forms the basis of the kinetic theory of gases which provides simple explanation of processes like diffusion, etc. However, experimentally, it is well known that ergodicity is broken, diffusion processes can have power law distributions of free path lengths such as Levy distributions, and so on.

Again, when dealing with long-range interactions, energy is not extensive. Moreover, in the standard classical statistical mechanics framework, it is not clear how to accommodate a multifractal structure for the relevant phase space, multifractal applications being in systems exhibiting scale invariance.

Fractals appear in chaotic dynamical systems and are associated with strange attractors. Fractals are unique because they have fractional dimensions. Different types of dimensions may be associated with fractals, for instance, the similarity dimension, the information dimension, the correlation dimension, etc.. Hentschel and Procaccia[11] carried out a study of generalized dimensions D_q, defined for a parameter $q \geq 0$ that characterized an attractor. In particular they found that for $q = 0$ the similarity dimension arises, information dimension for $q = 1$, correlation dimension for $q = 2$, and for integer values of $q = 3, 4, \ldots, n$ etc., the generalized dimensions are associated with correlation integrals of triplets, quadruplets,..and ntuplets of points on attractors. The information dimension is related to the of a system. Tsallis extended this idea to statistical mechanics.[12]

Historically, q-analogues and q-deformation of numbers have been of interest in mathematics since very long.[1-4] For instance, Heine[2] defined the deformation of a number n to a q-basic number as:

$$[n]_q = \frac{1 - q^n}{1 - q} \tag{1}$$

where $\lim_{q \to 1} [n]_q = n$. In fact, calculus, algebra, and all other mathematical operations and functions we are familiar with, can be deformed using a parameter q. Jackson[4] was the first to introduce the generalized derivative or the q-deformed derivative:

$$\partial_x^{(q)} f(x; y; \ldots) = \frac{f(qx; y; \ldots) - f(x; y; \ldots)}{(q - 1)x} \tag{2}$$

which measures the rate of change of the function with respect to a dilatation of its argument by a factor of q.

Using this q- deformation as a tool, many physical systems and some discrete nonlinear maps have been studied.[13-16] One possible explanation for the solar neutrino problem was enunciated by using a q deformed Gaussian as the distribution for the velocity of atoms in the sun.[17] The characteristic velocity of stars in a galaxy as predicted by theory was much higher than what was experimentally observed; introduction of q deformed variables made theoretical values closer to the experimental values. Similarly cold atoms in optical lattices appear to diffuse in a non-Gaussian pattern which correspond to certain values of $q \neq 1$.[18,19]

The concept of q-deformation may be motivated by considering the following differential equations. First, we observe that the differential equation $\frac{du}{dz} = u$ has the solution $\ln u = z$, or equivalently $u = e^z$. On the other hand, the differential equation

$$\frac{dy}{dx} = y^q \tag{3}$$

has the solution

$$y = [1 + x(1 - q)]^{1/(1-q)} \tag{4}$$

which upon inversion, gives

$$x = \frac{y^{1-q} - 1}{(1 - q)}. \tag{5}$$

$y = e_q^x$ and $x = \ln_q y$ are generalised solutions and have been defined as the deformed exponential and deformed logarithm functions respectively, in Tsallis' non-extensive statistical mechanics.[20] As $q \to 1$, the original exponential and logarithm functions are restored.

Using an expansion of e_q^x in a Taylor series around $x = 0$, another deformation of numbers was obtained[3]:

$$x_q = \frac{x}{1 + (1 - q)(1 - x)} = \frac{\frac{x}{2-q}}{1 + \frac{q-1}{2-q}x}. \tag{6}$$

In the discussion below, we show how the concept of q-deformation of numbers with Tsallis' statistics, as in eqn.(6), can be employed usefully in a mathematical model of an ecological situation which has been observed in nature and studied over several decades.

3. The Larch Bud-Moth Population Cycle

Observations spanning over more than a hundred years have noted outbreaks of the bud-moth insect on larch trees at high altitudes occurring at regular periodicity of 8-9 years in the Swiss Alps and in the North American continent. Large regions of luxuriantly green coniferous forests have been turned rapidly into brown, bearing a scorched appearance, because of the infestation by these insects and the larvae devouring upon the larch needles which subsequently dry up.[8] Larch bud-moth cycles are an example of herbivorous population cycles which are described by density-dependent changes in their food. These changes could arise, for instance, due to climactic conditions. As the population of bud-moth larvae increases, so does the infestation of the forest, thus leading to rapid reduction of the foliage available for the larvae to feed upon. This triggers a fall in the bud-moth numbers, giving the larch trees time to flourish again. The regrowth of larch needles is however arrested by the bud-moths again which indulge in wasteful destruction of the foliage. A stream of literature attributes the periodicity in the bud-moth population and in the needle lengths to the interaction between the herbivore and the larch needles alone. It may be borne in mind that forest trees, as prey, may fluctuate or cycle in the abundance of biomass or components such as nitrogen, rather than in numbers. Hence cycles of trees are often studied by using a quality that would be a representation of the health of the plants.

It is found that during heavy infestation, the protein content of the larch needle falls from 6 to 4 percent, while its fibre content increases from 12 to 18 percent. The average length of the needles is over 30mm, which during defoliation reduces to less than 20mm.

It turns out that models based on the assumption that plant quality alone affects the bud-moth's carrying capacity, have failed to capture the dynamics of the bud-moth population. This has come as a surprise as this implies that changes in food quality and bud-moth population are no longer first order as suspected.[22] Such models which do explain the short term effects of plant quality on the population of the bud-moth, also predict the delay in the bud-moth's population rise despite the increase in plant quality. In the observations made in the Swiss Alps, it was found that despite the very small reduction of leaf quality, the population of bud-moth severely went down.[22] It is now known that the plant quality affects the intrinsic growth rate and not its carrying capacity. This is indicative of the presence of other factors that might be affecting the bud-moth population.[22]

The Bud-Moth — Needle-Quality — Parasitoid Tritrophic Model

Since plant quality alone was not sufficient to explain the observed dynamics, a parasitoid-host interaction was added in to the system.[9] Although the larch bud-moth is attacked by as many as 94 parasitoid species,[23,24] parasitism rates at the bud-moth population peak are typically low — around 10–20 percent,[8] which indicates the fact that population of the larch bud-moths are not significantly controlled by the parasitoids. The parasitism rate however, reaches a high of around 80 percent during the collapse stage, i.e., when the population of the bud-moth is at the lowest. But again there is a phase lag between the peak of the parasitoid population and the peak of the bud-moth population. This model involving only the bud-moth and the parasitoids failed as it was found that bud-moth population fluctuations regulate the parasitoid population and not the other way round.[10] To overcome the difficulties faced by the above models, a tritrophic model was suggested[9,10] which included both parasitoid-interaction and plant quality influence on the bud-moth. This is in accordance with current understanding of the observed larch bud-moth cycles which points to primarily two factors affecting the population dynamics — the plant quality and the interaction with the parasitoids.[22] When one of them fails, the other one ensures that the nine-year cycle continues.

Larch needle lengths usually vary between 15 and 30mm. The length of the needle is a measure of the protein and fibre contents of the leaf. Lengthier needles are associated with lesser fibre and more protein content, and thus are good source of food for the bud-moths. A plant quality index (needle length or needle quality) of 1 denotes high quality food, while 0 corresponds to poor quality of food. This directly affects the female fecundity, larval growth and pupal biomass. The leaf quality at time t is denoted by L_t which may be made dimensionless by the following transformation[22]

$$Q_t = \frac{L_t - 15}{15}.$$

(7)

4. A q-deformed Tritrophic Model

The entire system can be dealt with as a q-deformed system. q-deformed systems are useful in situations where there is a memory, or strong correlation, or long distance neighbour interactions. The larch bud-moth population cycles have memory of previous years' growth in the system, and this system has a complex interaction with the parasitoids living on the bud-moth larvae, and the plant quality index. Hence it is highly motivating to look at the system as a q-deformed system, rather than as a simple ergodic system. In ergodic systems, the probability that the system is in any given state, is equal for all states. In a system where there is memory, however, the probability distribution is skewed. In other words, some probabilities are enhanced while some are suppressed. In such a scenario, use of standard physics fails. For instance, one can explain the movement of an amoeba in search of food using a q-Gaussian.[25,26] The probability distribution of searching is q-deformed and hence the system is no longer ergodic, and this is what is also observed in nature.

Hence we propose using Tsallis deformed variables to represent the parasitoid population, bud-moth population, and plant quality. In ecological systems, different organisms may respond differently. Hence it is natural to have different deformation parameters for different interacting organisms.

In this work, we introduce an additional interaction term between the bud-moth and the plant quality in the tritrophic model already extant in the literature,[9] and additionally, use deformed variables.

The three main variables of interest in the problem are the bud-moth population x_t, the parasitoid population y_t and the plant quality index z_t. We identify z_t with Q_t defined in eqn.(7). Using Tsallis deformation, as in eqn.(6), we then deform all these variables, each with a different deformation q_i, as follows:

$$x_q = \frac{\rho_x x}{1 + \mu_x x}, \quad y_q = \frac{\rho_y y}{1 + \mu_y y}, \quad z_q = \frac{\rho_z z}{1 + \mu_z z}, \tag{8}$$

where

$$\rho_i = \frac{1}{2 - q_i}, \quad \mu_i = \frac{q_i - 1}{2 - q_i}, \quad i = x, y, z. \tag{9}$$

We choose to work with $q_x = 1$, so that $x_q = x$.

Thus we consider the following system of equations with deformed variables $x_q = x$, y_q and z_q for modeling the larch bud-moth — needle quality — parasitoid dynamical system:

$$x_{t+1} = \lambda x_t z_{q_t} \exp(-x_t - y_{q_t})$$
$$y_{q_{t+1}} = c x_t [1 - \exp(y_{q_t})]$$
$$z_{q_{t+1}} = (1 - \alpha)(1 - \frac{x_t}{m + x_t}) - \alpha x_t z_{q_t}. \tag{10}$$

The first two equations describe the bud-moth and parasitoid through the Beddington model[27] type interactions, except that the prey (bud-moth) equation is linked (on the right-hand side) with z_q since needle quality directly affects bud-moth larval growth. The third equation describes the dynamics of the leaf quality. The quantity α may be related to memory of the previous growth as in Turchin's work,[10,22] and $1 - \alpha$ is the plant quality recovery rate. In the absence of the bud-moth, the leaf quality dynamics would be given by $Q_{t+1} = (1 - \alpha) + \alpha Q_t$. The decrement of the plant

quality because of its consumption by the bud-moth may be captured by
the multiplying factor $(1 - \frac{x_t}{m+x_t})$, m being the half-saturation constant
for uptake by the moth population. When the plant cannot regrow, the
plant quality decays to zero at the rate α. The last term in the equation
for the leaf quality incorporates effects of other direct interactions with the
bud-moth.

Our q-deformed model may be related to Turchin's tritrophic model
through the following transformations:

$$x_t = \beta N_t, \quad y_t = a P_t, \quad z_t = Q_t. \tag{11}$$

These scaling relations were actually used in[28] to rewrite Turchin's moth-
parasitoid model in dimensionless form — these authors however did not
consider the tritrophic model. Here N_t denotes the larch bud-moth densities at time t, λ denotes the bud-moth's intrinsic growth rate and P_t the
parasitoid population at time t. Turchin's tritrophic model comprises of
these equations:

$$N_{t+1} = \lambda N_t \frac{Q_t}{\delta + Q_t} \exp\left(-\beta N_t - a\frac{P_t}{1 + awP_t}\right)$$

$$P_{t+1} = b N_t \left[1 - \exp\left(-a\frac{P_t}{1 + awP_t}\right)\right]$$

$$Q_{t+1} = (1 - \alpha)\left(1 - \frac{N_t}{\gamma + N_t}\right) + \alpha Q_t \tag{12}$$

where

$$\beta = \exp\left[\frac{1}{K}\ln\left(\frac{N_{t+1}}{N_t}\right)\right]. \tag{13}$$

Here a denotes the searching rate and w the wasting time for the parasitoids.
The number of surviving parasitoids coming from each parasitized moth is
denoted by b; the bud-moth carrying capacity is denoted by K; δ denotes the
half saturation constant for plant quality, and γ denotes the half saturation
constant for moth population uptake upon the plant.

From eqn.(13), we see that β is an indicator of the intensity of competition within the bud-moth population, for resources and space.

The constants m and c in eqn.(10) can be related to the constants in
Turchin's equations through the relations: $m = \beta\gamma$, $c = \frac{ba}{\beta}$.

Comparing Eqs. (10) and (12), and using eqns.(8) and (9), we see that it
is possible to lend physical interpretations to the deformation parameters
q_y and q_z which can take various values depending upon the scenario
under consideration. The deformed variable y_q can be identified with the

parasitoid functional response $\frac{aP_t}{1+awP_t}$ which incorporates the fact that a finite time is spent by the predators (the parasitoids) in searching for the prey (bud-moths) and that not all the prey are hunted or consumed.[29]

μ_y can take values from 0 to ∞. μ_y is related to the parasitoid wasting time w: $\mu_y = \frac{q_y - 1}{2 - q_y} = w$. The time wasted by the predator in hunting the prey can theoretically vary from 0 to ∞. The value zero corresponds to the situation in which the predator is continuously consuming the prey, without any time gap between two encounters. Similarly $\mu_y = \infty$ corresponds to a situation when the organisms wastes infinite amount of time in searching for prey, this implies that the organism would starve to death before finding food to eat or when there is no prey available for consumption. Also, we must bear in mind that w is positive. Thus, q_y varies between 1 and 2.

μ_z is related to the plant quality half saturation constant δ: $\mu_z = \frac{1}{\delta}$, and $\rho_z = \frac{1}{\beta\delta}$. These give: $q_z = 1 + \beta$. The intraspecific competition coefficient β can theoretically vary from 0 to ∞, the value zero corresponding to a situation where there is no competition between the larch bud-moth for resources or space. This would happen when the bud-moth population is very low or available resources are very high. Similarly, $\beta = \infty$ would correspond to a situation when there is excessive competition between the bud-moths for food and space — this would happen when the population level is very high or when the available resources are very low. Since δ is a positive quantity, it limits the value of q_z to vary between 1 and 2.

Linear Stability Analysis

The Jacobian of the system is given by

$$
\begin{pmatrix}
\lambda z_q e^{-x-y_q} & \frac{-x\lambda\rho_y e^{-x-y_q}}{(1+\mu_y y)^2} & \frac{\lambda x \rho_y e^{-x-y_q}}{(1+\mu_z z)^2} \\[2mm]
c(1 - e^{-y_q}) & -c\frac{\rho_y e^{-y_q}}{(1+\mu_y y)^2} & 0 \\[2mm]
\frac{(\alpha-1)m}{(m+x)^2} + \alpha z_q & 0 & \frac{\alpha x \rho_z}{(1+\mu_z z)^2}
\end{pmatrix}.
$$

A fixed point of the system is $(0, 0, (1-\alpha))$; other fixed points are the roots of equations (10). The Jacobian of the system at this fixed point is

$$
\begin{pmatrix}
\lambda(1 - \alpha) & 0 & 0 \\[2mm]
0 & -c\rho_y & 0 \\[2mm]
\frac{\alpha-1}{m} + \frac{\alpha(1-\alpha)\rho_z}{1+(1-\alpha)\mu_z} & 0 & 0
\end{pmatrix}.
$$

For the same fixed point, the eigenvalues Λ are 0, $-c\rho_y$, $\lambda(1-\alpha)$. For maps, if $|\Lambda| > 1$, then the fixed point is unstable. For this fixed point $(0,0,(1-\alpha))$ to be stable, we require that $|-c\rho_y| < 1$, $|\lambda(1-\alpha)| < 1$. α varies between 0 and 1, hence, $0 \leq (1-\alpha) \leq 1$. Therefore the fixed point $(0,0,(1-\alpha))$ would require $\lambda < \frac{1}{1-\alpha}$ for stability. Similarly, we have from the other eigenvalue that $c < \frac{1}{\rho_y}$, with $1 \leq \rho_y \leq 2$ for the fixed point to be stable. Since λ quantifies the intrinsic growth of the bud-moth population, as its value increases, the fixed point would become more and more unstable. Again, for a given value of c, ρ_y may increase in such a way that $|\Lambda| > 1$, destabilizing the fixed point. An increase in ρ_y would imply a decrease of q_y, which translates to the situation in which the parasitoid is not wasting much time in hunting down its prey. That is, the parasitoid becomes more efficient and its population begins to grow. $\Lambda = 0$ suggests the existence of an unstable manifold.

As parameter values change, there are also other periodicities such as 8.25, or 7.071 etc., which occur in our model. Nature, however, strongly adheres to the 8-9 year cycle.

Dynamics of the System for Various Parameter Values

The larch bud-moth cycle is known for its regular periodicity of about 8-9 years. Our q-deformed model reproduces this periodicity in certain parameter regimes.

Our system has 6 parameters; we vary one of them at a time, keeping the others constant and study the bifurcation diagrams. Bifurcation diagrams give a qualitative understanding of the change in the equilibria of the system as the parameters of the system vary. A system's periodic or chaotic behavior can be inferred from these.

The plots in Fig. (1) depict the power spectrum of the data obtained after numerically solving our q-deformed equations (10). We find that all the three variables in the system have a dominant 9 year period cycle. Along with the 9 year cycle there is also a smaller cycle for 4.5 years, which is rather prominent for the parasitoids.

The time series of the bud-moth population, parasitoid population and plant quality index, calculated from our deformed equations (eqns.(10)), show that they are all mutually synchronized in activity though separated mutually by constant phase differences of small magnitudes (as observed also in nature).

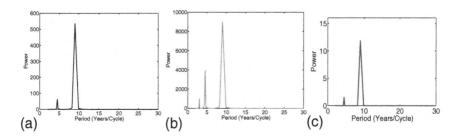

Fig. 1. Power spectrum of: (a) bud-moth population, (b) parasitoid population and (c) plant quality index. Parameter values are: $c = 10$, $\alpha = 0.5$, $u = 0.4$, $m = 15$, $\lambda = 12.8125$ and $q = 1.18$. The power spectrum shows the 9 year cycle observed in all three quantities.

In Fig. (2), bifurcation diagrams of the three variables are generated as a function of the deformation parameter q_z, keeping q_y at a fixed value. Although the three variables show transition to chaotic and periodic regimes at the same values of q_z, we find that bud-moth population and parasitoid population undergo crises at $q_z = 0.47$ and $q_z = 0.52$ respectively, while the needle length undergoes boundary crisis at $q_z = 0.47$ and internal crisis at $q_z = 0.52$. Since there is sudden burst of stable points, we expect the existence of an internal crisis, i.e., an initially chaotic attractor having contacted an unstable manifold of a saddle, makes the entire unstable manifold an attractor. Hence the attractor size increases. The needle length also shortens as its predator (bud-moth) population increases. Parasitoid population depends upon that of the bud-moth and hence it increases as bud-moth population increases.

In Fig. (3), the populations of bud-moth and parasitoid, and the plant quality index are plotted against the various values of the deformation parameter q_y, for a given value of q_z. As expected, the needle length goes down as the bud-moth population increases. At $q_y = 1.18$, a 9 period cycle is seen in all the variables — bud-moth population, parasitoid population and the plant quality index. Since q_y is related to the wasting time w, as q increases, the wasting time increases and the influence of the parasitoids on the bud-moth decreases. The bud-moths increase plant consumption, thereby reducing plant quality index. However this excessive intake of leaf by the bud-moths reduces their available resources to feed upon, thereby

S. V. Iyengar and J. Balakrishnan

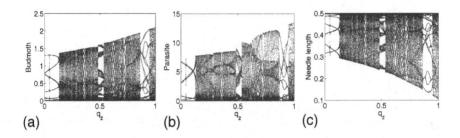

(a)   (b)   (c)

Fig. 2. Bifurcation diagrams for variation with respect to q_z for: (a) Bud-moth population, (b) Parasitoid population and (c) plant quality index. Parameter values are: $\alpha = 0.5, m = 15, c = 10, \lambda = 12.8125$ and $q_y = 1.18$.

reducing their population. Consequently, the leaf quality increases again. This is reflected in Figs. 3(a-c).

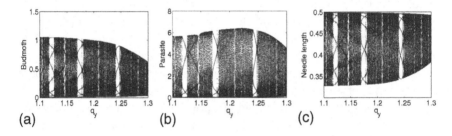

(a)   (b)   (c)

Fig. 3. Bifurcation diagrams for variation with respect to q_y for: (a) Bud-moth population, (b) Parasitoid population and (c) plant quality index. Parameter values are: $\alpha = 0.5, m = 15, c = 10, \lambda = 12.8125$ and $q_z = 0.4$.

In Fig. (4), α, the parameter related to memory is varied and the responses of the bud-moths, parasitoids and needle-lengths are studied. For $\alpha > 0.55$ we find the occurrence of boundary crisis. A boundary crisis occurs when a stable attractor contacts an unstable manifold of a saddle and ceases to exist. Boundary crises are associated with the decrease in the size of the attractor.[30] An attractor is a region in space that attracts all the initial conditions towards it. An attractor can exist for certain values of parameters and completely disappear for some other values.

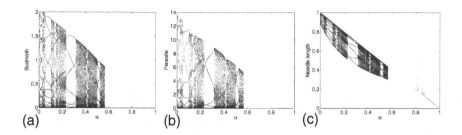

Fig. 4. Bifurcation diagrams for variation with respect to α for: (a) Bud-moth population, (b) Parasitoid population and (c) plant quality index. Parameter values are: $q_y = 1.18, m = 15, c = 10, \lambda = 12.8125$ and $q_z = 0.4$.

Figure (5) depicts the bifurcation diagrams of two variables simultaneously as the deformation parameter q_z is varied — in (a) the bud-moth population and the plant quality (needle) index simultaneously, and in (b) the parasitoid population and the plant quality index simultaneously — we observe the richness in the dynamical behavior possible: there are transitions from chaotic to periodic to multiply-periodic states in different parameter regimes. Period halving transitions are also seen.

5. Conclusions

The deformation parameters q_y and q_z are respectively related to the parasitoid wasting time w and the intraspecific competition coefficient β for the bud-moth population. As the deformation q_y increases, it physically corresponds to parasitoid wasting time increasing. This implies that the parasitoids are no more efficient in attacking their prey and hence their population decline for higher values of q_y. Because of this, the bud moths survive, feeding excessively on leaves, making the needle length shorter. This is also seen in the bifurcation diagrams. Due to the unavailability of high quality food the bud moth population also goes down.

As the competition between bud-moths for resources decreases, q_z increases. Thus the bud-moth population increases and hence the parasitoid population increases also. This is seen in the bifurcation diagrams (Figs. (2) & (3)). Since the bud-moth population increases, the needle length goes down (Fig. (4)).

λ, the intrinsic growth rate of the bud-moth can change very easily, as growth rates can always be associated with changes in temperature,

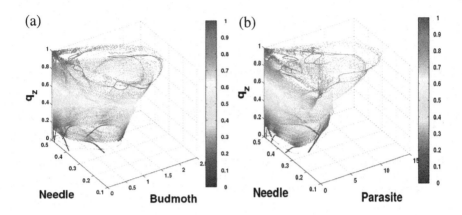

Fig. 5. Bifurcation diagrams showing variation with respect to q_z for (a) both bud-moth population and plant quality index, and (b) both parasitoid population and plant quality index. Parameter values are: $\alpha = 0.5, m = 15, c = 10, \lambda = 12.8125$ and $q_y = 1.18$. The color code indicates values of q_z.

climate, etc. q_y which is related to the parasitoid wasting time can change if the predator is not efficient — for instance, if they are drugged (through use of pesticides). Since the population of bud-moths, parasitoids and the plant quality index vary, and q_z is related to the total bud-moth population at any given time, q_z can also vary.

The larch bud-moth population cycle has been having a regular periodicity of about 8–9 years, since many, many years. It is intriguing that mathematical models are so sensitive to changes in parameter variations; yet at least 50 years of well-documented data available, show remarkable accuracy of the 8–9 year cycle. Perhaps there are more parameters in the system that keep this 8–9 year cycle very stable, and which are yet to be identified.

While the bifurcation diagrams in this present work cover extensively all the parameter regimes of our modified and improved q-deformed tritrophic model, we are not aware of a similar exercise having been carried out for the original tritrophic model.

The parasitoid cycle in our model shows a significant 4.5 year cycle in additional to the original dominant 9 year cycle. There are also other sub-harmonics which are not significant. All the three variables i.e., the bud-moth population, the number of host parasitoids and the leaf quality index show period doubling route to chaos, or period-halving from a chaotic behavior as the parameter is varied (Figs. (2–4), for $q_z > 0.8$).

The period halving or period doubling happens in all the three populations at exactly at the same values of the parameters. This indicates the fact that they influence each other strongly. The leaf quality index reaches a maximum of 1, in the absence of either parasitoid or bud-moth, or both. This is also a fixed point of the system. This confirms the known fact that in absence of the bud-moths, the plant can grow well. In our q-deformed model, the larch bud-moth and parasitoid populations show boundary crisis with respect to the memory parameter α, and internal crisis with respect to the deformation parameter q_z in a certain parameter regime.

This study illustrates the complex connections between different trophic levels in a small ecosystem. The q-deformation parameters quantify in different ways, information transfer across various levels of biological organization. The magnitude of large-scale rapid destruction and defoliation of vast areas of forest cover can well be governed by seemingly insignificant quantities like the parasitoid wasting time or the intraspecific competition coefficient for the herbivore population. The nature of the nonlinearities in the mathematical model reproduce the cycles which are actually observed in nature and indicate the possibility of gaining control over the system's behaviour, getting it to operate in a "desirable" parameter regime.

6. Acknowledgement

S.V. Iyengar acknowledges financial support from the Council of Scientific and Industrial Research, New Delhi.

References

1. L. Euler, *Introduction in Analysis Infinitorum*, Busquet, Lausanne (1748).
2. E. Heine, *Handbuch der Kugelfunktionen*, Vol.1, Reimer, Berlin (1878), reprinted by Physica-Verlag, Wurzburg (1961).
3. F.H. Jackson, *Proc. Roy. Soc. London* **74**, 64 (1904).
4. F.H. Jackson, *Trans. Roy Soc. Edin.* **46**, 253 (1908).
5. C. Tsallis, F. Baldovin, R. Cerbino and P. Pierobon, "Introduction to nonextensive statistical mechanics and thermodynamics" in *The Physics of Complex Systems: New Advances and Perspectives; Proceedings of the International School Of Physics 'Enrico Fermi'*, **155**, p. 229, eds. F. Mallamace & H. E. Stanley, IOS Press (2004).
6. A.A. Berryman, "Population Cycles: Causes and Analysis", in *Population cycles: The case of trophic interactions*, ed. A. Berryman, pp. 3–28, Oxford University Press (2002).

7. R.A. Werner, "Biology and behavior of a larch bud moth, Zeiraphera sp., in Alaska". Res. Note PNW-RN-356. Portland, OR: U.S. Department of Agriculture, Forest Service, Pacific Northwest Forest and Range Experiment Station. 8 pages (1980).

8. W. Baltensweiler and A. Fischlin, "The Larch Budmoth in the Alps", in *Dynamics of Forest Insect Populations*, ed. A. Berryman, pp. 331–351, Plenum Publishing Corporation (1988).

9. P. Turchin, S.N. Wood, S.P. Ellner, B.E. Kendall, W.W. Murdoch, A. Fischlin, J. Casas, E. McCauley and C.J. Briggs, *Ecology*, **84**, 1207 (2003).

10. P. Turchin, C.J. Briggs, S.P. Ellner, A. Fischlin, B.E. Kendall, E. McCauley, W.W. Murdoch, and S.N. Wood, "Population Cycles of the Larch Budmoth in Switzerland", in *Population cycles: The case of trophic interactions*, ed. A. Berryman, pp. 130–141, Oxford University Press (2002).

11. H.G.E. Hentschel and I. Procaccia *Physica D* **8**, 435 (1983).

12. C. Tsallis, *J. Stat. Phys.* **52**, 479 (1988).

13. R. Jaganathan and S. Sinha, *Phys. Lett. A* **338**, 277287 (2005).

14. F. Brouers and O. Sotolongo-Costa, *Physica A* **368**, 165 (2006).

15. V. Patidar and K.K. Sud, *Commun. Nonlin. Sci. Numer. Simul.* **14**, 827 (2009).

16. S. Banerjee and R. Parthasarathy, *J. Phys. A: Math. Theor.* **44**, 045104 (2011).

17. G. Kaniadakis, A. Lavagno and P. Quarati, *Phys. Lett. B* **369**, 308 (1996).

18. E. Lutz, *Phys. Rev. A* **67**, 051402(R) (2003).

19. P. Douglas, S. Bergamini and F. Renzoni, *Phys. Rev. Lett.* **96**, 110601 (2006).

20. C. Tsallis, *Quimica Nova* **17**, 468 (1994).

21. E.P. Borges, *J. Phys. A* **31**, 5281 (1998).

22. P. Turchin, *Complex Population Dynamics: A Theoretical/Empirical Synthesis*, Princeton University Press (2003).

23. A.A. Berryman, *Trends Ecol. Evol.*, **11**, 28 (1996).

24. V. Delucchi, *Entomophaga*, **27**, 77 (1982).

25. F.L. Schuster and M. Levandowsky, *J. Euk. Microbiol.*, **43**, 150 (1996).

26. G.M. Viswanathan, S.V. Buldyrev, S. Havlin, M.G.E. da Luz, E.P. Rapaso and H.E. Stanley, *Nature* **401**, 911 (1999).

27. J.R. Beddington, C.A. Free, and J.H. Lawton, *Nature* **255**, 58 (1975).

28. S. R.-J. Jang and D. M. Johnson, *J.Biol.Dyn.* **3**, 209 (2009).

29. N. J. Gotelli, *A Primer of Ecology*, Sinauer Associates, Inc. (2001).

30. C. Grebogi, E. Ott, and J. A. Yorke, *Science*, **238**, 632 (1987).

Chapter 9

Newtonian chimpanzees? A molecular dynamics approach to understanding decision-making by wild chimpanzees

Matthew Westley[1], Surajit Sen[1] and Anindya Sinha[2,3]

[1] *Department of Physics, State University of New York at Buffalo,*
Buffalo, New York 14260-1500, USA

[2] *National Institute of Advanced Studies, Indian Institute of Science Campus,*
Bangalore 560012, India;
asinha@nias.iisc.ernet.in

[3] *Nature Conservation Foundation, 3076/5 4th Cross, Gokulam Park,*
Mysore 570002, India

In this study, we computationally investigate decision-making by individuals and the ensuing social structure of a primate species, chimpanzees, using Newton's equations of classical mechanics, as opposed to agent-based analyses in which individual chimpanzees make independent decisions. Our model uses molecular dynamics simulation techniques to solve Newton's equations and is able to approximate the movements of female and male chimpanzees, especially in relation to the available food resources, in a manner that is consistent with their observed behaviour in natural habitats. It is noteworthy that our Newtonian dynamics-based model may allow us to make certain specific observations of their behaviour, some of which may be difficult to achieve through agent-based modelling exercises or even field studies. Chimpanzees tend to live in fission-fusion social groups, with varying number of individuals, in which both females and males tend to display intrasexual competition for valuable food resources while the males also compete for oestrus females. Most populations of the species are also restricted to a small range of habitats, a clear indication that they are especially vulnerable to the availability and distribution of food sources. With reasonable assumptions of chimpanzee behaviour, we have been able to analyse the clustering behaviour of individuals in relation to local food sources as also patterns of their migration across groups. Our simulated results are qualitatively consistent with field observations conducted on a particular semi-isolated population of chimpanzees in Bossou, Guinea, in western Africa.

1. Introduction

Cognitively driven social decisions made by individuals primarily determine the structure and dynamics of animal social groups, particularly those of human and nonhuman primates (see, for example, Refs. 1–3). Such decision-making is usually strongly influenced by a variety of factors that confer certain advantages to individual members of these groups, including defense against predation, exclusive access to resources and, consequently, prolonged survival and increased reproductive success.[4–7] These benefits should, however, outweigh the occasional concomitant costs, such as increased competition over food, which socially aggregated individuals face in order for such associations to be stable. The availability of food, both in terms of its quantity and distribution (both spatial and temporal), and the ease with which group-living animals may be able to detect and harvest such food resources constitute one of the most important forces that drive individual animals to associate with one another, either over their lifetimes or temporarily, during certain times of the year (see, for example, Refs. 8 and 9).

In nonhuman primates, ecological and social pressures have together been invoked to explain not only the formation of societies, but also the size of social groups and the pattern of interactions between individuals within and across such groups.[7,10] Thus, according to the prevailing socioecological models, ecological factors appear to be most crucial in determining the spatial distribution and social relationships of group-living female primates, with such relationships usually reflecting a trade-off between the potential for feeding competition, which is a function of the quantity and distribution of resources and group size,[11–13] and anti-predator benefits.[14] In contrast, male social relationships and ranging are directly affected by the distribution and fertility of females over space and time,[7,15] and the optimisation of these sex-specific competitive strategies shape the social structure of many primate species.[6,7] While such patterns have indeed been empirically established for several species of female-bonded primates (as, for example, blue monkeys,[16] or bonnet macaques,[3] grouping decisions can be significantly more complex for other primate species.

A classic example of a species with complex socioecological decision-making is the chimpanzee, evolutionarily our closest nonhuman relative.

Chimpanzees live in a variety of habitats, from savanna woodlands and woodland-dry forest mosaics to rainforests.[17] They feed mainly on ripe fruit but also consume leaves, other plant parts, insects, and occasionally cooperatively hunt monkeys and other mammals.[18] Given the critical requirement for ripe fruit, chimpanzee ranges invariably include some proportion of evergreen or riverine forest, even when they reside in semiarid regions. Moreover, most chimpanzee groups require large home ranges, of the order of $5-20$ km^2 per community in Gombe National Park in Tanzania[19] and up to $30-40$ km^2 at other forest sites,[20] primarily because their food resources are patchy and widely dispersed. Chimpanzees living in drier habitats with more spatially and temporally scattered food resources usually require much larger home ranges.[21]

Akin to humans, chimpanzees are unusual amongst mammals in displaying male philopatry whereby males tend to remain in their natal groups while females typically disperse from these groups (termed "communities") upon reaching sexual maturity.[22] Moreover, unlike many other group-living primates, chimpanzees (*Pan troglodytes*) exhibit a fission−fusion grouping pattern, with community members associating in temporal groups called "parties", which vary in size, composition and duration in an unpredictable manner.[22,25] Male chimpanzees appear to form highly differentiated social bonds with dyads that not only preferentially associate in parties but also affiliate and pa cooperate in other behavioural contexts.[26] Although female chimpanzees tend to affiliate and cooperate less with one another than do males,[27-31] they do form stable associations within parties and affiliate strongly with one another in some populations.[32-36] The stability of intrasexual party associations also does not differ between the two sexes over the years.[33]

Various factors have been proposed to explain the variability in the size of these parties and their ranging patterns, including food availability and distribution, travel costs, predator pressure, demography, social bonding and affiliation, number of estrous females in the group, within-group aggression, and behavioural traditions.[22,25,37-40] Investigations into the relationship between party size, individual movement and food availability revealed that changes in the availability and distribution of fruits, the principal food resource for chimpanzees, was possibly the most crucial factor determining grouping and group size in this species[37,41,42]; but see.[43]

An important consequence of grouping in chimpanzees is escalated conflict, especially over food resources. As a result of flexible grouping, however, chimpanzees can easily avoid direct conflict over food by traveling in smaller parties or alone when food becomes scarce.[22] In their association patterns, anestrous females appear particularly sensitive to food availability: they tend to join parties at rich food sources and move away from them when food is depleted.[28] Although dominance interactions are less frequent among females than among the males of this species, superior access to food may contribute to greater reproductive success among females.[44] In general, females are believed to be less gregarious than males ([28,45,46]; but see above) and spend much of their time alone.[28,47] Thus, in the face of feeding competition, the strategy of female chimpanzees would rather be to disperse and track scattered food resources rather than to compete with one another over clumped food.[43,48]

Chimpanzee males tend to travel farther in a day and to range more widely than do females.[4,28] At Gombe in Tanzania, for example, males increased their day range when food availability was high.[49] In addition to food availability, the availability of estrous females, predation or certain social factors such as the presence of stranger males also influence the day range and party size of male chimpanzees in many habitats.[23,25,41,49–51]

Theoretical approaches to the study of animal grouping patterns have had a long history. The different methods that have been used include qualitative[52] and mathematical arguments[53–74] as well as numerical simulations.[68,70,75–82] The general consensus that emerges is that the clustering behaviour common to many species can be modelled by a pair of distance-dependent attractive and repulsive forces, resulting in a stable spacing between animals within the group.[52,83,84] It is noteworthy that such spacing between individuals is well supported by classic field observations for several species in nature.[54,85–87]

Two of the most common approaches to modelling grouping behaviour are the Eulerian (continuum) and Lagrangian (discrete) methods, which are generally used to study very large or very small groups of animals.[74] Our study is essentially Lagrangian in nature: we consider animals as individual particles in a simulation subject to attractive and repulsive forces. The biological motivation for the precise form of these forces has been supported either by field observations of animal grouping patterns (as, for example,[88] or appropriate models of individual behaviour[89–91]).

In this paper, we wish to present a possible numerical recipe for Lagrangian modelling up to the mesoscale. Agent-based and cellular automata models have frequently been used for this purpose. These approaches are useful as the biological motivations for the agent behaviours/ cellular automata rules are intuitive; agents representing prey, for example, should flee from predators. After the parameters for each type of agent have been defined, one would place a large number of these agents in a region and allow them to interact with one another for a period of time and then explore for any large-scale emergent behaviours such as clustering or long-range interactions that may emerge (see[92] for a recent review). These models are, however, often computationally intensive for large numbers of animals and mesoscale studies below the range of applicability of Eulerian models can, therefore, be difficult. Furthermore, it is a difficult task to determine an analytical form for the resulting behaviour from these models. Given these limitations, we decided to adopt a different approach – the well-known technique of molecular dynamics (see[93]) – to address the problem. It is noteworthy that such an approach is based on common interactions between chimpanzees could generate the collective behaviour we seek. We have, of course, as alluded to above, focussed on the specific case of a single species, chimpanzees, in this paper although this problem can and has been earlier studied in a broader context.[94,95]

As Newton's laws can be applied, rather usefully, to the grouping dynamics of certain animal populations, we have taken advantage of pre-existing molecular dynamics models and programs that have already been highly optimised and are openly accessible. We then use the model that we develop to determine two quantities of interest: the degree of clustering of a given group of animals and the migration rate between two groups. It should be mentioned here that as there is only limited field data on chimpanzees for several reasons, including the small number of individuals observed and the often inaccessible nature of their habitats, we compare our model to field observations whenever possible.

2. Results

We begin by imagining several types of particles representing individual animals and their food sources. In a typical molecular dynamics study, we would then determine which atomic interactions ("potentials") would be most important. Here, we may imagine that individual animals will have a long-range attraction towards food sources and a short-range repulsion towards one other. The long-range attraction is intended to be a simple model of whichever modality is used by the individuals to detect and acquire food while the repulsion represents a minimisation of competition that is typically observed in nature for most species ([83,96] including chimpanzees.[22,48,97] To model each species then, we only need to modify the exact shape of these interaction potentials – perhaps one species has better abilities or mechanisms to detect food from a greater distance or is perhaps less tolerant toward competition and can thus be modelled by a stronger repulsion force. The main problem with this approach comes from our adherence to Newton's laws of motion: an "equal and opposite reaction" may be appropriate when considering two identical individuals but not necessarily for an asymmetric relationship, as, for example, a predator-prey relationship.[98] This style of modelling, therefore, will be best suited to interactions within a single species or between multiple species competing for similar resources.

This paper focuses on a single species, examining intra- and inter-group interactions of chimpanzees, especially in relation to the distribution of food resources (see[42] for the complex relationship between group patterns and food availability in this species). For a single species and a single food source, the basic experimental setup is simple. We create a large region (10,000 km²) and place two types of particles uniformly randomly about the region. The first type of particle is rare, perhaps 10−20 in the entire region. These are fixed to a single location, and represent an area that contains a food resource. The second type of particle represents individual chimpanzees; they move under the effects of the interaction potentials described above (attraction to food, repulsion from one another). Further details of these interaction potentials are discussed in Appendix A. We then allow this system to evolve in time according to Newton's laws of motion for a sufficiently long time such that an equilibrium is reached. As discussed in Appendix A, the "long time" is approximately between 2 and 6 months. This equilibrium is what we are particularly interested in, as this is the state that an observer would be most likely to observe during field studies on the species.

The particular functions used for the interaction potentials are shown below. The animal-animal interaction, $V_A(r)$, is a Lennard−Jones (LJ) 12−6 potential (so called because of the powers of r in the denominator), with parameters ϵ, σ and a cut-off range (Eq. 1):

$$V_A(r) = 4\epsilon \left[\left(\frac{\sigma}{r}\right)^{12} - \left(\frac{\sigma}{r}\right)^{6} \right] r < r_{c1}. \tag{1}$$

The animal-food interaction $V_F(r)$ is a Yukawa function with a positive exponent and parameters A, k and r_c (Eq. 2):

$$V_F(r) = A\frac{e^{kr}}{r} r < r_c. \tag{2}$$

Fig. 1 depicts a plot of these functions. Eq. (2) implies that it is challenging for an individual animal to obtain food from distant sites. Thus, this equation influences how individuals cluster near food sources and we discuss this behaviour in detail below. It is important to note here that, given the relative inaccessibility of chimpanzee habitats across their range from western to eastern Africa, it may be quite difficult to observe the actual effects of changing food resources on social interactions and decision-making in chimpanzees (but see Hashimoto *et al.* 2003, Chapman *et al.* 1995;[42,48] and Potts *et al.* 2011). The dynamics of such processes can, however, be inferred from the molecular dynamics simulations that we explore here.

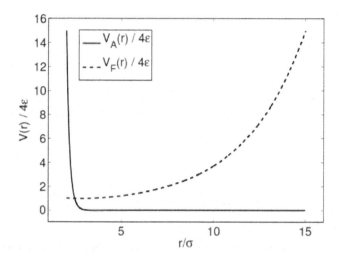

Fig. 1.   Short-range repulsion between individuals $V_A(r)$ and long-range attraction between individuals and food resources $V_F(r)$. The forces have been divided by 4ϵ or σ to make the quantities unitless, as described in Appendix A.

We usually observe one of three types of behaviour: (I) a state resembling ordinary thermal equilibrium, where the animals are finally uniformly distributed, (II) the state shown in Fig. 2a – a high degree of clustering near the food sources, and finally, (III) a brief transition phase between the earlier two states, where the individuals are clearly not uniformly distributed, but there is no real boundary between the clusters.

This clustering behaviour is extremely sensitive to changes in the interaction parameters, in particular k and σ (see further details in Appendix A). The parameter σ is effectively the amount of "personal space" each individual requires, the distance at which two individuals begin to repel one another. A slight increase in this parameter ($\sim$0.1 km) tends to break up the clusters, forcing the system into State I. k has a similarly significant effect; it directly modifies the strength of the attraction of individuals towards the food sources. Increasing k by 10% approximately doubles the

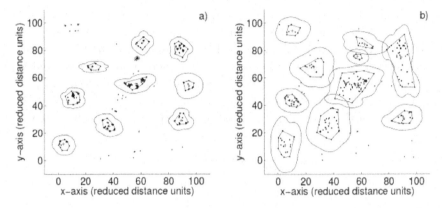

Fig. 2. Equilibrium positions of the particles. The placement of food sources is identical, and so the clusters form in similar locations. Two calculations of the home range area are shown: the inner line is a minimum convex polygon drawn around the cluster, and the outer line comes from a kernel density estimate. (a) Low-temperature run: $T^* = 0.6, r_C^* = 10$; (b) High-temperature run: $T^* = 2.0, r_C^* = 10$.

strength of the food-animal attraction, causing all the individuals to cluster as tightly as possible. This high sensitivity is an indication that the parameters we have chosen are, in fact, useful. In order to analyse the exact relationship between such clustering and the interaction parameters, we must define some measure for cluster size. There are several options here – we may measure the area of the cluster (for example, by the minimum convex polygon method) or the average separation between individual points. These options are discussed in Appendix B. Fig. 2 shows two different measurements of group area while Figs. 3a and 3b depicts measurements of cluster size using the length-based method described in Appendix B, which we denote by R ($R^* \equiv R/\sigma$ is a unitless distance). As R increases, the clusters become more spread out over space. Fig. 3a plots R vs. the simulation temperature (which equates to the mobility of the species), and shows a linear increase with temperature. The relationship between R and the force is more complex. Fig. 3b shows R vs. the attractive force's cutoff; we manipulate the parameters r_c and k such that the range of the force increases while keeping its maximum value constant. Refer to Fig. 1; this is simply stretching the Yukawa attractive potential out horizontally while keeping the maximum height invariant. Fig. 3b shows a power law decay in the size of the cluster. The two area-based methods thus agree with these results, but the data is much noisier.

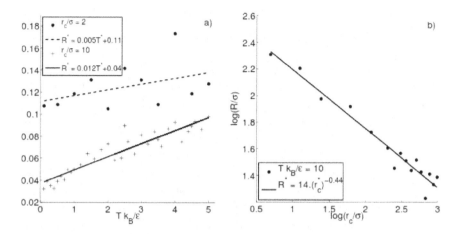

Fig. 3. The clusters become more tightly packed as the potential cutoff increases, and temperature decreases. (a) R vs. temperature for two fixed values of r_C, with linear fit; (b) R vs. r_C for fixed T, with power-law fit.

It may be reiterated here that our model is now capable of examining the effect of changes in food resources on core areas defended by chimpanzee groups as also grouping patterns that could be displayed by the species. We believe that this model could also be used to predict the impact of changes in the availability and distribution of food resources, effected by anthropogenic changes such as climate change, deforestation, or the spread of agriculture and farming, on the already endangered populations of chimpanzees in the future.

It should also be mentioned here that the reduced distance units of the simulation may be intuitively viewed as kilometers. The units of time will change with the potential as time is a derived unit (unlike mass, length, and energy in LJ reduced units) and, for this reason, we usually consider the state of the groups only after sufficient time has elapsed for the system to relax into an equilibrium state. We discuss the time units further below.

The molecular dynamics framework is very robust and can potentially be used to model more complex social structures, brought about, for example, by adding more types of particles to the simulations. To cite a specific example, we note that the "personal space" parameter has a strong effect on group formation; it could thus be of interest to add a small number of individuals (say, aggressive dominant males) that have a slightly different σ from the rest of the individuals. As another example, we could place an entire ecosystem of individual animals in the model by using several types of particles that have significantly different types of interactions with one another.

We have so far discussed the structure of our molecular dynamics model and some test results. But is the model useful in explaining patterns of chimpanzee distribution and movement encountered in the field? As stated above, the strength of the molecular dynamics approach is in studying spatial structure; it may thus be illustrative to examine the structure of chimpanzee groups as established by field observations.

Chimpanzees are known to exhibit high rates of inter-group migration.[25,46,99] This species is unusual amongst primates in displaying extensive female transfer, with a majority (50 to 90%) of females leaving their natal troops, after attainment of sexual maturity, to associate and form coalitions with unrelated females in other groups.[25,100,101] Female chimpanzees, which often travel alone, disperse widely and form small ranges.[100] They are usually selective in their choice of groups and prefer to join units with relatively greater number of males.[102] Male chimpanzees, in contrast, tend to remain in their natal groups throughout their life and often form kin-based coalitions in order to maintain their social status.[23,103,104] The typical fission-fusion features of chimpanzee groups, described above, are usually shaped by male mating strategies and female dispersal to avoid competition for food (Wrangham 1986). Chimpanzee populations, thus, usually consist of several unit-groups that females transfer into, often against aggression displayed by the resident females, and which are strongly defended by males against adjacent groups or communities.[7,5,25,100] Aggression between stranger males or groups can be severe with lone males driven off, injured, or even killed.[105,106] Within each of these unit-groups, males usually employ variable reproductive strategies while females form mating relationships with different males.[25]

In an unexpected variation, however, a 21-year study of a semi-isolated chimpanzee group at Bossou, Guinea found a surprisingly large number of males immigrating into and emigrating out of the group; a migration rate "not less than that of females" (Sugiyama 1999). Sugiyama hypothesizes that the absence of competing neighbouring groups could render male alliances relatively unimportant in this population while the limited availability of food resources and females could promote elevated intra-group competition among the males — both factors leading to increased male movement. Moreover, the existence of a food resource concentrated in a small core area within the region may stabilise the overall population size in Bossou.[99]

We have tested Sugiyama's hypothesis using our model. First, we added the ability of the model to distinguish between males and females and, furthermore, between the individuals of two initially separate groups. In molecular dynamics terms, this is as simple as running the simulation with five particle types (males and females of Group 1, males and females of Group 2, and the food source) rather than with two (individuals and food), as was done initially. Let us call the first four new particle types M1, M2, F1, and F2 respectively. We then adjust the interactions between these particles appropriately. The hostility between strange males is handled by doubling the interaction energy between M1 and M2 while the attraction between males and foreign females is modelled by decreasing the interaction energy between M1 and F2, and between M2 and F1. Finally, we place the food source in the centre of these two groups, and test the effect of food localization on the migration rates of individuals. The localization of food is handled by changing the slope of the attractive potential; a steeper increase to the maximum attraction represents, in effect, a localised food supply. The model is otherwise left identical.

We then run the simulation, averaging over 40 runs, and record the independent migration of M1, M2, F1, and F2 between the two groups as a function of time. This is plotted in Fig. 4. The first panel shows the number of M1 and F1 within Group 1 as a function of time or, effectively, the number of individuals that do not migrate from this group. The females have a slightly higher migration rate, as would be expected from our knowledge of chimpanzee social structure being primarily driven by female transfer. The second panel shows exactly the same data, but for a weakly localised food source.

From these plots, we observe that the effect of strong localisation of the food resource is to reduce the number of males in the group; many of the males leave the group to join the other group or wander alone. This is also what would be predicted from Sugiyama's observation of male chimpanzee migration in Bossou and his hypothesis that this results from the strong competition between the males for a localised food resource (and hence, localised females as well) in this isolated population.[99] We have thus been able to establish the qualitative validity of the model and may now be able to use it to make qualitative predictions of migration patterns across these chimpanzee groups as the availability and distribution of the food resource changes over time.

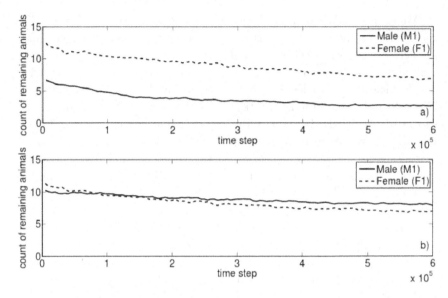

Fig. 4. The number of individuals that remain within their natal group as a function of the simulation time step. The decay is only caused by migration, as the total number of individuals is fixed. (a) Strongly localised food supply; (b) Weakly localised food supply.

Table 1. The approximate rate of individual male and female chimpanzees in three age classes remaining in their respective natal groups in Bossou, Guinea (adapted from Sugiyama 1999)

| | Age Class | | |
Sex	Infant (0 to 4 years)	Juvenile (4 to 8 years)	Adolescent (8 to 12 years)
Females	0.80	0.80	0.33
Males	0.80	0.64	0.14

Now, the conversion to real-world time units becomes important when our model is being applied to real-time data, as we are no longer considering only the equilibrium state of the system. As mentioned above, the time scale depends on the potential, which can be determined in this particular example by comparing it to the known migration rates. Sugiyama measured the rate of individual males and females in three age classes (infant, juvenile and adolescent) remaining in their respective natal groups[99] and this is shown in Table 1. Our simulations do not reveal the significant change in the rate of migration observed by Sugiyama across each four-year age class, as our model does not incorporate any qualitative (biological) change in its constituent particles over time. We can, nevertheless, at least determine an approximate time scale for the simulations; a year is about 10^6 simulation time steps.

3. Conclusions

It is important to note that this simulation is designed to capture the features of available data on chimpanzee clustering with a very simple-minded approach; we, however, do have the ability to add more complex features to the model without significantly increasing the computation time. The method was a simple Newtonian molecular dynamics-based approach, treating chimpanzees and their food sources as particles interacting under short- and long-range potentials. The observation of clustering in the data is not very surprising (indeed it is the first milestone that we worked towards), but the high degree of sensitivity to the shape of the potential functions is promising. The particular potentials that we have used led to the appearance of temperature and force cutoff as important parameters in the model − these would equate to the mobility and attraction range (for the food source) of the chimpanzees in their natural habitats. It would also be important to examine how, with the addition of different types of functions and multiple species of particles, it may be possible for us to model the behaviour of complex social groups without adding a priori too much detail into the model's structure. This is particularly important, as our model has, so far, only been able to predict the apparently unusual migratory behaviour of the Bossou chimpanzees.[99] Given the sensitivity of the observed dynamics to the model parameters, however, we should be able to parameterise the model suitably in order to explain and predict the more usual behavioural patterns of chimpanzee groups across their distribution range. These patterns are, of course, typically shaped by the sex-specific decisions made by female chimpanzees to transfer from their natal groups to other groups and of males to be philopatric and develop coalitions with related individuals in their natal groups. More generally, such models allow us to explore the nature of ecological and social information acquired by individuals in complex societies and which, in turn, influence the cognitive decisions made by individuals in course of their life-history strategies.

Finally, it must be mentioned here that, in recent years, the already spatially limited habitats of chimpanzees are under severe threat from various anthropogenic factors such as deforestation, habitat fragmentation and the excessive harvesting of timber and other forest resources.[19,107–109] This, coupled with the increasing effects of global climate change, is likely to wreak havoc with the abundance and distribution of potential food resources for most chimpanzee populations.[109] It is thus becoming imperative to predict what the future holds for these threatened chimpanzee groups as their living space and food sources continue to shrink indomitably over the years. It may thus be important to develop our model further so that it can predict long-term changes in chimpanzee populations via a grand canonical ensemble that allows for the creation or annihilation of its constituent particles under defined conditions.

4. Appendix A: The Model, in Detail

The molecular dynamics suite we use is known as LAMMPS [Large-scale Atomic/Molecular Massively Parallel Simulator].[110] To set up the simulation, we create a 100 by 100 region in dimensionless length units, and place 240 animals and 12 food sources randomly scattered in the region. The numbers, of course, may be varied. We then define the potential in terms of pair interactions. For this, we use two different potential functions, as the interaction between individuals is fundamentally different from that between the individuals and their food (assuming, of course, naturally omnivorous chimpanzees to be herbivorous for the purposes of our model). Note that the form of the potential is not very restrictive; we just require an attractive potential to model a food source. Recall that the potential functions, plotted in Fig. 1, are

$$V_A(r) = 4\epsilon \left[\left(\frac{\sigma}{r}\right)^{12} - \left(\frac{\sigma}{r}\right)^6 \right] r < r_{c1}$$

and

$$V_F(r) = A\frac{e^{kr}}{r} r < r_c.$$

This particular form increases with distance, which is intended to simulate the growing need for food when an individual is starving. The food particles are stationary in order to represent vegetation or fruit sources as food; this is implemented simply by making the mass of the food particles large, as compared to other masses in the simulation. The particular values of $k = 0.50106$ and $r^*C = 10$, used in Fig. 2, are selected such that the food-animal and animal-animal potentials are of the same order of magnitude; increasing k by 0.05 will approximately double the strength of the food-animal attraction, causing all of the animals to cluster as tightly as possible while decreasing it by the same amount will approximately halve the strength of the attraction, destroying the clusters.

LAMMPS has the built-in capability to make all quantities dimensionless ("reduced Lennard-Jones units"), which we generally use, as many important physical quantities such as mass or energy may not have direct analogues when modelling animals. We set the time step to 0.0005 of the reduced unit time, defined by LAMMPS. We will later see that in this simulation, the unit time is of the order of several hours, so the time step is only a few minutes long. The units in the figures are dimensionless unless stated otherwise. We use the notations T, r etc. to represent units with dimension and T^*, r^*, etc. and to represent reduced units (while $T^* = Tk_B/\epsilon, r^* = r/\sigma$, etc. have been used in the standard way). This system of units has been defined such that the reduced parameters are near unity, whether dealing with atomic / molecular systems or that of our own.

The terms r_{C1} and r_C are the force cutoffs for the LJ and Yukawa potentials respectively. The temperature, we believe, should be just as important in this model as it is in a normal molecular dynamics simulation. Physically, it is a measure of mean kinetic energy. When dealing with animals, then, it seems reasonable to surmise that the temperature corresponds to a measure of mobility or, in other words, how much individuals tend to move in search of food. This is not a quantity usually measured in theoretical models, but we believe that "temperature" is still a useful quantity when applied to animal populations. It should be possible to evaluate this measure in the field through certain techniques such as radio-telemetry. With these parameters defined, we start a single run of the simulation with a long-time NVE time integration (a micro-canonical ensemble, in the statistical mechanics language[111] to allow the system to relax from its initial randomly scattered state). This takes only about 50,000 time steps, which is roughly equivalent to teo months, due to the small number of particles, but we generally triple that number for a safety margin. After the system has relaxed, we switch to the NVT integrator (a canonical ensemble,[111] using the Nose-Hoover equations of motion[112–114] with a heat bath at temperature T, in the form given by Ref. 115). The time integration routine is a

PPPM solver built in to LAMMPS, described in.[116] Energy is conserved to about 1% with a time step of 0.0005 during the 150,000 time steps of the relaxation phase, which can be improved by reducing the time step. For our purposes, 1% energy conservation, combined with many runs to reduce errors from "drifting" energy, was found to be adequate.

The parameters and values described here correspond to a single species with an experimentally determined average mobility. If we wish to add another species with a different mobility, we would just have to change the mass from the reduced mass of 1 that all the animals currently have. Moreover, by allowing different masses for particles that represent different categories of individuals such as males, females, dominant or subordinate individuals and the like, we can easily add more detail to the social interactions that require to be modeled. We, however, did not undertake this exercise at the first stage of our modelling, as we wanted to model the relatively simple behavioural clusters displayed and the decisions made by the chimpanzees of the Bossou population.

We have chosen to study mainly the spatial distribution of our study animals with this approach, as the computational methods of molecular dynamics do not lend themselves well to population models. After some experimentation, we have found that it is not easy to build an ensemble that can handle births and deaths of individuals. We have thus created an alternative structure, similar to a grand canonical ensemble. Unfortunately, it is difficult to define a chemical potential equivalent that can handle birth/death phenomena correctly and in fact, LAMMPS itself is not designed to handle changes in particle number. Our model is, therefore, not well suited to handling birth and death processes. What this approach can do rather efficiently instead is to rapidly and easily show how the animals will tend to arrange themselves after enough time has elapsed for the system to relax. Using the potential functions defined above, we can make the animals interact with one another and with their food sources in appropriate ways. We have, thus, so far only studied the potential shown in Eq. (1) and (2). For each simulation run, we allow the system sufficient time to reach equilibrium and then extract the final positions of the animals. An example using the parameters $\sigma = 0.5, \epsilon = 4, r^*_{C1} = 2, A = 1, k = 0.50106, r^*_C = 10$ and $T^* = 0.6$ is shown in Fig. 1. (Recall that the asterisk denotes reduced units.) The exact value of the LJ cut-off, r^*_{C1} is not important, as the potential dies off to nearly zero at that distance.

5. Appendix B: Group Analysis

There are two problems that need to be solved in order to study the structure of groups: first, we must partition the particles into a certain number of groups, and second, we must invoke specific representation for the spatial extent of the clusters.

The partitioning can be done with any one of many standard techniques of cluster analysis.[117] One of the most commonly used is the k-means algorithm.[122,123] This method, though with an advantage of simplicity, does not handle outliers well. We thus use a MATLAB implementation of the single-linkage hierarchical clustering algorithm.[119] The partitioning can also, of course, be conducted by eye (see Fig. 2) but this becomes a difficult exercise when handling large data sets.

Once the particles have been separated into clusters, we must determine a measure for the size of each cluster. We may measure the area of a cluster by estimating the home range of the study animals (as the particle clusters, in this case, simulate groups of animals). This is done most commonly by drawing the smallest convex polygon (with no internal angle greater than 180°), which contains each of the points in a cluster. This estimate, however, tends to be subject to systematic bias and other methods are usually preferred.[120] Kernel density estimation is a more reliable, though complex, method. Essentially, the probability distribution function for the point clusters is numerically approximated and the smallest contour of this function that contains, say, 90% of the points is chosen as the home range. Fig. 2 shows an example of both these methods.

These measures of group size are both noisy estimates. We, therefore, also use a third approach, which is length-based rather than area-based. For a given cluster of points, we calculate the mean distance to the centroid of the cluster (the group average dissimilarity.[121] With identical initial conditions we expect this distance to be the same for every cluster in a simulation (run up to some noise) and thus use the mean of these distances as a single indicator of the size of the clusters. This number is plotted in Fig. 3 for various initial conditions.

Acknowledgements

Matthew Westley and Surajit Sen are grateful to the National Science Foundation URGE Program, USA for support of this work. Anindya Sinha would like to thank B V Sreekantan and Janaki Balakrishnan for their kind invitation to participate in this volume and for their patient enthusiasm!

References

1. A. Sinha, *Philos. T. Roy. Soc. B* **353**, 619 (1998).
2. A. Sinha, *J. Biosci.* **30**, 51 (2005).
3. A. Sinha, K. Mukhopadhyay, A. Datta-Roy and S. Ram, *Curr. Sci.* **89**, 1166 (2005).
4. R. W. Wrangham, *Soc. Sci. Inform.* **18**, 336 (1979).
5. R. W. Wrangham, *Behaviour* **75**, 262 (1980).
6. C. P. van Schaik and J. van Hooff, *Behaviour* **85**, 1 (1983).
7. R. I. M. Dunbar, *Primate social systems*, Crook Helm, London (1988).
8. S. Altmann, *Am. Zool.* **14**, 221 (1974).
9. J. W. Bradbury and S. L. Vehrencamp, *Behav. Ecol. Sociobiol.* **1**, 383 (1977).
10. B. Smuts, D. L. Cheney, R. M. Seyfarth, R. W. Wrangham and T. T. Struhsaker (eds.) *Primate societies*, University of Chicago Press, Chicago (1987).
11. L. A. Isbell, *Behav. Ecol.* **2**, 143 (1991).
12. E. H. M. Sterck, D. P. Watts and C. P. van Schaik, *Behav. Ecol. Sociobiol.* **41**, 291 (1997).
13. A. Koenig, *Int. J. Primatol.* **23**, 759 (2002).
14. C. P. van Schaik, *The ecology of social relationships amongst female primates*, in V. Standen and R. Foley (eds), *Comparative socioecology*, Blackwell, London (1989), pp. 195–218.
15. C. P. van Schaik, *Proc. Brit. Acad.* **88**, 9 (1996).
16. T. M. Butynski, *Ecol. Monogr.* **60**, 1 (1990).
17. G. Teleki, *Population status of wild chimpanzees (Pan troglodytes) and threats to survival*, in P. G. Heltne and L. A. Marquardt (eds.), *Understanding chimpanzees*, Harvard University Press, Cambridge, Massachusetts (1989), pp. 312–353.
18. T. Nishida, R. W. Wrangham, J. Goodall and S. Uehara, *J. Hum. Evol.* **12**, 467 (1983).
19. A. E. Pusey, G. W. Oehlert, J. M. Williams and J. Goodall, *Int. J. Primatol.* **26**, 3 (2005).
20. M. L. Wilson and R. W. Wrangham, *Ann. Rev. Anthropol.* **32**, 363 (2003).
21. P. J. Baldwin, W. C. McGrew and C. E. G. Tutin, *Int. J. Primatol.* **3**, 367 (1982).
22. J. C. Mitani, D. P. Watts and J. S. Lwanga, *Ecological and social correlates of chimpanzee party size and composition*, in C. Boesch, G. Hohmann, & L. Marchant (eds.), *Behavioural diversity in chimpanzees and bonobos*, Cambridge University Press, Cambridge (2002), pp. 102–111.
23. T. Nishida, *Primates* **9**, 167 (1968).
24. T. Nishida and K. Kawanaka, *Kyoto Univ. Afr. Stud.* **7**, 131 (1972).
25. J. Goodall, *The chimpanzees of Gombe — patterns of behavior*, Harvard University Press, Cambridge, Massachusetts (1986).

26. J. C. Mitani, D. A. Merriwether and C. B. Zhang, *Anim. Behav.* **59**, 885 (2000).
27. S. Halperin, *Temporary association patterns in free ranging chimpanzees: an assessment of individual grouping preferences*, in D. Hamburg and E. Mc-Cown (eds.), *The great apes*, Benjamin/Cummings, Menlo Park, California (1979), pp. 491–499.
28. R. W. Wrangham and B. B. Smuts, *J. Rep. Fer. S.* **28**, 13 (1980).
29. K. Kawanaka, *Int. J. Primatol.* **5**, 411 (1984).
30. T. Hasegawa, *Sex differences in ranging patterns*, in T. Nishida (ed.), *The chimpanzees of the Mahale Mountains: sexual and life history strategies*, University of Tokyo Press, Tokyo (1990), pp. 99–114.
31. J. M. Williams, A. E. Pusey, J. V. Carlis, B. P. Farm and J. Goodall, *Anim. Behav.* **63**, 347 (2002).
32. J. W. Pepper, J. C. Mitani and D. P. Watts, *Int. J. Primatol.* **20**, 613 (1999).
33. I. C. Gilby and R. W. Wrangham, *Behav. Ecol. Sociobiol.* **62**, 1831 (2008).
34. J. Lehmann and C. Boesch, *Int. J. Primatol.* **29**, 65 (2008).
35. K. Langergraber, J. C. Mitani and L. Vigilant, *Am. J. Primatol.* **71**, 840 (2009).
36. M. Wakefield, *Anim. Behav.* **85**, 1303 (2013).
37. C. A. Chapman, R. W. Wrangham and L. J. Chapman, *Behav. Ecol. Sociobiol.* **36**, 59 (1995).
38. D. Doran, *Int. J. Primatol.* **18**, 183 (1997).
39. J. C. Mitani and S. J. Amsler, *Behaviour* **140**, 869 (2003).
40. J. C. Mitani, *Primates* **47**, 6 (2006).
41. R. W. Wrangham, C. A. Chapman, A. P. Clark-Arcadi and G. Isabirye-Basuta, *Social ecology of Kanyawara chimpanzees: implications for understanding the costs of great ape groups*, in W. C. McGrew, L. F. Marchant and T. Nishida (eds.), *Great ape societies*, Cambridge University Press, Cambridge (1996), pp. 45–57.
42. N. Itoh and T. Nishida, *Primates* **48**, 87 (2007).
43. M. Wakefield, *Int. J. Primatol.* **29**, 907 (2008).
44. M. Emery Thompson, S. M. Kahlenberg, I. C. Gilby and R. W. Wrangham, *Anim. Behav.* **73**, 501 (2007).
45. M. Emery Thompson and R. W. Wrangham, *Comparison of sex differences in gregariousness in fission-fusion species: reducing bias by standardizing for party size*, in N. E. Newton-Fisher, H. Notman, V. Reynolds and J. Paterson (eds.), *Primates of Western Uganda*, Springer, New York (2006), pp. 209–226.
46. R. W. Wrangham, *Why are male chimpanzees more gregarious than mothers? A scramble competition hypothesis*, in P. M. Kappeler (ed.), *Primate males: causes and consequences of variation in group composition*, Cambridge University Press, Cambridge (2000), pp. 248–258.
47. C. M. Murray, L. E. Eberly and A. E. Pusey, *Behav. Ecol.* **17**, 1020 (2006).
48. S. M. Kahlenberg, M. Emery Thompson and R. W. Wrangham, *Int. J. Primatol.* **29**, 931 (2008).

49. R. W. Wrangham, *Feeding behaviour of chimpanzees in Gombe National Park, Tanzania*, in T. H. Clutton-Brock (ed.), *Primate Ecology*, Academic Press, London (1977), pp. 503–537.
50. C. E. G. Tutin, W. C. McGrew and P. J. Baldwin, *Primates* **24**, 154 (1983).
51. C. Boesch, *Behaviour* **117**, 220 (1991).
52. A. E. Parr, *Occas. Pap. Bingham Oceanogr. Coll.* **1** (1927).
53. C. M. Breder, *B. Am. Mus. Nat. Hist.* **98**, 3 (1951).
54. C. M. Breder, *Ecology* **35**, 361 (1954).
55. D. S. Cohen and J. D. Murray, *J. Math. Biol.* **12**, 237 (1981).
56. T. Nagai and M. Mimura, *SIAM J. Appl. Math.* **43**, 449 (1983).
57. T. Nagai and M. Mimura, *J. Math. Soc. Jpn.* **35**, 539 (1983).
58. A. Okubo, *Adv. Biophys.* **22**, 1 (1986).
59. P. Grindrod, *J. Math. Biol.* **26**, 651 (1988).
60. S. Gueron and N. Liron, *J. Math. Biol.* **27**, 595 (1989).
61. P. Turchin, *Comments Theor. Biol.* **1**, 65 (1989).
62. D. Grünbaum, *J. Math. Biol.* **33**, 139 (1994).
63. D. Grünbaum and A. Okubo, *Modelling social animal aggregations*, in S. A. Levin (ed.), *Frontiers in mathematical biology*, Lecture notes in biomathematics, Vol 100. Springer-Verlag, Berlin (1994), pp. 296–325.
64. E. E. Holmes, M. A. Lewis, J. E. Banks and R. R. Veit, *Ecology* **75**, 17 (1994).
65. A. Mogilner and L. Edelstein-Keshet, *Physica D* **89**, 346 (1996).
66. A. Mogilner, L. Edelstein-Keshet and G. Bard Ermentrout, *J. Math. Biol.* **34**, 811 (1996).
67. A. Mogilner and L. Edelstein-Keshet, *J. Math. Biol.* **38**, 534 (1999).
68. A. Okubo and S. A. Levin, *Diffusion and ecological problems*, Springer, New York (2002).
69. V. Gazi and K. M. Passino, *IEEE Transactions on Automatic Control* **48**, 692 (2003).
70. D. Morale, V. Capasso and K. Oelschlger, *J. Math. Biol.* **50**, 49 (2005).
71. M. Burger, V. Capasso and D. Morale, *Nonlinear Anal-Real.* **8**, 939 (2007).
72. M. Burger and M. Di Francesco, *Netw. Heterog. Media* **3**, 749 (2008).
73. V. Capasso and D. Morale, *Stoch. Anal. Appl.* **27**, 574 (2009).
74. V. Capasso and D. Morale, *On the stochastic modelling of interacting populations. A multiscale approach leading to hybrid models*, in P. J. Neittaanmaki (ed), *Applied and numerical partial differential equations: scientific computing in simulation, optimization and control in a multidisciplinary context*, Springer (2010), pp. 59–80.
75. D. Morale, *Appl. Math. Comput.* **10**, 157 (2000).
76. D. Morale, *Future Gener. Comp. Sy.* **17**, 883 (2001).
77. S. H. Lee, H. K. Pak and T. S. Chon, *J. Theor. Biol.* **240**, 250 (2006).
78. B. Nabet, N. E. Leonard, I. Couzin and S. Levin, *Leadership in animal group motion: a bifurcation analysis*, in *Proceedings of the 17th International Symposium on Mathematical Theory of Networks and Systems*, Kyoto, Japan (2006), pp. 1–14.

79. S. J. Moon, B. Nabet, N. E. Leonard, S. A. Levin and T. G. Kevrekidis, *J. Theor. Biol.* **246**, 100 (2007).
80. J. Lu, J. Liu, I. D. Couzin and S. A. Levin, *Emerging collective behaviors of animal groups*, in *Intelligent Control and Automation, 2008*, WCICA 2008. 7th World Congress, pp. 1060–1065.
81. B. Nabet, N. E. Leonard, I. Couzin and S. A. Levin, *J. Nonlinear Sci.* **19**, 399 (2009).
82. A. Eriksson, M. N. Jacobi, J. Nystrom and K. Tunstrøm, *Behav. Ecol.* **21**, 1106 (2010).
83. J. A. Beecham and K. D. Farnsworth, *J. Theor. Biol.* **198**, 533 (1999).
84. A. Mogilner, L. Edelstein-Keshet, L. Bent and A. Spiros, *J. Math. Biol.* **47**, 353 (2003).
85. P. Conder, K. Williamson, E. Stheskmann and F.Hemming, *Ibis* **93**, 315 (1951).
86. J. T. Emlen, *Auk* **69**, 160 (1952).
87. R. S. Miller and W. J. D. Stephen, *Ecology* **47**, 323 (1966).
88. A. S. Glen and C. R. Dickman, *Biol. Rev.* **80**, 387 (2007).
89. S. Gueron, S. A. Levin and D. I. Rubenstein, *J. Theor. Biol.* **182**, 85 (1996).
90. S. V. Viscido, M. Miller and D. S. Wethey, *J. Theor. Biol.* **208**, 315 (2001).
91. S. V. Viscido, M. Miller and D. S. Wethey, *J. Theor. Biol.* **217**, 183 (2002).
92. C. Castellano, S. Fortunato and V. Loreto, *Rev. Mod. Phys.* **81**, 591 (2009).
93. M. P. Allen and D. J. Tildesley, *Computer simulation of liquids*, Oxford University Press, New York (1989).
94. I. Couzin, J. Krause, R. James, G. Ruxton and N. Franks, *J. Theor. Biol.* **218**, 1 (2002).
95. A. Wood and G. Ackland, *P. Roy. Soc. B-Biol. Sci.* **274**, 1637 (2007).
96. J. Krause and G. Ruxton, *Living in groups*, Oxford University Press, New York (2002).
97. J. Riedel, M. Franz and C. Boesch, *Am. J. Primatol.* **71**, 1 (2010).
98. J. Fryxell, A. Mosser, A. Sinclair and C. Packer, *Nature* **449**, 1041 (2007).
99. Y. Sugiyama, *Primates* **40**, 61 (1999).
100. T. Nishida, *The social structure of chimpanzees of the Mahale Mountains*, in D. A. Hamburg and E. R. McCown (eds.), *The great apes*, Benjamin/Cummings, Menlo Park, California (1979), pp. 73–121.
101. T. Furuichi, *Int. J. Primatol.* **10**, 173 (1989).
102. T. Nishida, M. Hiraiwa-Hasegawa, T. Hasegawa and Y. Takahata, *Z. Tierpsychol.* **67**, 284 (1985).
103. J. Goodall, *Z. Tierpsychol.* **61**, 1 (1983).
104. M. Hiraiwa-Hasegawa, T. Hasegawa and T. Nishida, *Primates* **25**, 401 (1984).
105. S. M. Kahlenberg, M. Emery Thompson, M. N. Muller and R. W. Wrangham, *Anim. Behav.* **76**, 1497 (2008).
106. R. Wrangham and L. Glowacki, *Hum. Nat.* **23**, 5 (2012).
107. P. Marchesi, N. Marchesi, B. Fruth and C. Boesch, *Primates* **36**, 591 (1995).
108. A. E. Pusey, L. Pintea, M. L. Wilson, S. Kamenya and J. Goodall, *Cons. Biol.* **21**, 623 (2007).

109. J. F. Oates, *Primates* **47**, 102 (2006).
110. S. Plimpton, *J. Comput. Phys.* **117**, 1 (1995).
111. C. Thompson, *Classical equilibrium statistical mechanics*, Clarendon Press, Oxford (1988).
112. S. Nose, *J. Chem. Phys.* **81**, 511 (1984).
113. S. Nose, *Mol. Phys.* **52**, 255 (1984).
114. W. G. Hoover, *Phys. Rev. A* **31**, 1695 (1985).
115. W. Shinoda, M. Shiga and M. Mikami, *Phys. Rev. B* **69**, 134103 (2004).
116. S. Plimpton, R. Pollock and M. Stevens, *Particle-mesh ewald and rRESPA for parallel molecular dynamics simulations*, in *Proceedings of the Eighth SIAM Conference on Parallel Processing for Scientific Computing*, Minneapolis, USA (1997), pp. 8–21.
117. M. Anderberg, *Cluster analysis for applications*, Probability and mathematical statistics, Vol 19, Academic Press, New York (1973).
118. J. MacQueen, *Some methods for classification and analysis of multivariate observations*, in L. M. Le Cam and J. Neyman (eds.), *Proceedings of the fifth Berkeley symposium on mathematical statistics and probability*, University of California Press, Berkeley, pp. 281–297.
119. R. Sibson, *R Comput. J.* **16**, 30 (1973).
120. M. Burgman and J. Fox, *Anim. Conserv.* **6**, 19 (2003).
121. T. Hastie, R. Tibshirani and J. Friedman, *The elements of statistical learning: data mining, inference, and prediction*, Springer, New York (2009).
122. C. Hashimoto, S. Suzuki, Y. Takenoshita, J. Yamagiwa, A. K. Basabose and T. Furuichi, *Primates* **44**, 77 (2003).
123. K. B. Potts, D. P. Watts and R. W. Wrangham, *Int. J. Primatol.* **32**, 669 (2011).

Chapter 10

Quantum Probability — A New Direction for Modeling in Cognitive Science

Sisir Roy

Physics and Applied Mathematics Unit, Indian Statistical Institute,
203 B.T. Road, Kolkata 700 108, India
sisir.sisirroy@gmail.com

Human cognition is still a puzzling issue in research and its appropriate modeling. It depends on how the brain behaves at that particular instance and identifies and responds to a signal among myriads of noises that are present in the surroundings (called external noise) as well as in the neurons themselves (called internal noise). Thus it is not surprising to assume that the functionality consists of various uncertainties, possibly a mixture of aleatory and epistemic uncertainties. It is also possible that a complicated pathway consisting of both types of uncertainties in continuum play a major role in human cognition. For more than 200 years mathematicians and philosophers have been using probability theory to describe human cognition. Recently in several experiments with human subjects, violation of traditional probability theory has been clearly revealed in plenty of cases. Literature survey clearly suggests that classical probability theory fails to model human cognition beyond a certain limit. While the Bayesian approach may seem to be a promising candidate to this problem, the complete success story of Bayesian methodology is yet to be written. The major problem seems to be the presence of epistemic uncertainty and its effect on cognition at any given time. Moreover the stochasticity in the model arises due to the unknown path or trajectory (definite state of mind at each time point), a person is following. To this end a generalized version of probability theory borrowing ideas from quantum mechanics may be a plausible approach. A superposition state in quantum theory permits a person to be in an indefinite state at each point of time. Such an indefinite state allows all the states to have the potential to be expressed at each moment. Thus a superposition state appears to be able to represent better, the uncertainty, ambiguity or conflict experienced by a person at any moment demonstrating that mental states follow quantum mechanics during perception and cognition of ambiguous figures.

1. Introduction

Decision making is one of the basic cognitive processes of human behaviors by which a preferred option or a course of actions is chosen from among a set of alternatives based on certain criteria. Historically, discussions of the science underlying decision making goes back to the 1600s — to Descartes and Poincare who created the first 'calculus' of decision making. The 20th century decision making approach goes back to Herbert Simon. Decision making is one of the fundamental cognitive processes of human beings that is widely used in multiple disciplines such as cognitive informatics, cognitive science, computer science, psychology, management science, decision science, economics, sociology, political science, and statistics. Each of those disciplines has laid emphasis on a special aspect of decision making. The cognitive perspective of decision making stands apart from the classical "game theoretic" perspective in which decision making is cast as a problem in optimization in terms of uncertainties, utilities and costs of the alternatives. From a "cognitive" perspective, decision-making is a dynamic process in which temporally varying noisy information is integrated across multiple time scales with a decision resulting when the information stream relevant to one of the alternative actions crosses a threshold. While the game theoretic approach will identify the optimum decision, the cognitive approach has a neurological basis but is often biased depending on the pattern in which the information is received. Thus, a cognitive perspective is the most realistic, if not essential, approach when studying animal and human decision-making in dynamic, real-time environments.

It is recognized that there is a need to seek an axiomatic and rigorous model of the cognitive decision-making process in the brain, which may serve as the foundation of various decision making theories. Decision theories can be categorized into two paradigms: the descriptive and normative theories. The former is based on empirical observation and on experimental studies of choice behaviors; and the latter assumes a rational decision-maker who follows well-defined preferences that obey certain axioms of rational behaviors.

Recently, several groups of scientists categorized the empirical findings [1] in various experiments related to decision making in the cognitive domain into six broad classes. We will discuss these six broad categories in human decision making, and the inadequecy of classical paradigms like classical probability theory will be discussed. The quantum paradigm will be discussed, where the concept of quantum probability will be discussed to explain the findings consistently. This will lead us to introduce new ontology called quantum ontology for understanding human cognition.

2. New Empirical Evidences

Empirical findings in various experiments related to decision making in the cognitive domain have been categorized into the following six classes:

- Disjunction effect
- Categorization — decision interaction
- Perception of ambiguous figures
- Conjunction and Disjunction fallacies
- Overextension of Category membership
- Memory recognition over-distribution effect and fallacies

We briefly discuss these categories following:[1]

- Disjunction effect: Amos Tversky and Eldar Shafir[2] discovered a phenomenon called the disjunction effect (Tversky and Shafir, 1992) in the process of testing a rational axiom of decision theory called the sure thing principle[3] (Savage, 1954). As per the sure thing principle, if action A is preferred over action B under a state of the world X, and this preference continues under the complementary state of the world X_C, then this preference of action A over B even when in ignorance of the state of the world. This was experimentally tested by Tversky and Shafir as follows: 98 students were presented with a two-stage gamble (a gamble which can be played twice). The decision to be made at each step was whether or not to go ahead and accept a gamble, there being an equal chance of either gaining 200$ or losing 100$. The key results could hardly be explained by simply considering either a win or loss. Busemeyer *et al.*[4] explained the findings to be an interference effect, in analogy to the results of double slit experiments done in particle physics. In both the situations there are two possible paths. The disjunction experiment requires an inference of the outcome with the first gamble of either a win or a loss. The two paths in a double split experiment correspond to the photon-splitting into two channels by a beam splitter. Both experiments have either knowledge or ignorance of the path taken, and in both cases in the unknown (unobserved) condition, the probability (be it gambling for the disjunction experiment, or of detection at a location for the two slit experiment) is much lesser than each of the probabilities for the known cases. Hence in the disjunction experiment, we may guess that under the unknown condition, the player instead of definitely being in one or the other state (win or loss), is instead in a superposed state. This makes it impossible to give a reason for choice of a gamble.

- Categorization — decision interaction: Several authors[5] studied the interaction between categorization and decision making, which gives rise to a new paradigm. This leads one to test Markov models and Quantum models.[6] Each trial involved showing participants pictures of faces, with a variation along two dimensions (face width and lip thickness). The participants were then required to categorize the faces as belonging to one of two groups – either "good" or "bad", and/or decide whether to make one of two actions: "attack" or "withdraw". Each participant was presented with two test conditions (but on different trials). One was the "C-then-D" condition, with participants making a categorization followed by an action decision. The second was the "D-alone" condition, involving only an action decision from the participants. There was a total of 26 participants, all undergraduate students from a Midwest US university. Each person participated for 6 groups of C-D trials, there being 34 trials per group, and one group of D-alone trials with 34 trials. One would think that as per the law of total probability the attack probability in D-alone condition would be a weighted average of the two probabilities of attacking conditioned on each categorization state. However, one sees that the total probability in the C-then-D trials is greatly exceeded by the probability of attack for D-alone. Moreover, the D-alone attack probability even exceeds the attack probability after categorization of a face as "bad"!

- Perception of ambiguous figures: Conte *et al.*[7] first investigated the interference effects in the perceptual domain by studying ambiguous figures. Participants were randomly divided between two groups. Members of one group were given a time of 3 seconds to make a single binary choice (either "plus" or "minus") with respect to an ambiguous figure A. The other group had the same time to make a single binary choice for an ambiguous figure B, but then followed 800 milliseconds later by a 3-second presentation requesting another single binary choice (plus vs. minus) for figure A. Significant interference effects were then observed. When test A was preceded by test B, it was seen that: $(P(B+)) :=$ probability of plus to figure B= 0.62, $P(A+|B+) :=$ probability of plus to figure A given plus to figure B= 0.78, $P_T(A+) :=$ total probability of plus to figure A= 0.69, $P(A+) :=$ probability of plus to figure A alone $= 0.55$. The interference effect corresponds to the difference $P_T(A+) - P(A) = +0.14$, which is not negligible and cannot be explained by the law of classical probabilities. So an alternative approach is called for to model this phenomenon.

- Conjunction and Disjunction fallacies: Tversky and Kahneman[8] studied the probability error called the conjunction fallacy in the following manner: Judges are provided a story (e.g., about a woman, Linda, earlier a philosophy student at a liberal university and an active anti-nuclear activist), and then asked to rank the likelihood of the following events: event F (e.g., Linda is now active in the feminist movement), event B (e.g., Linda is now working as a bank teller) and event $F \cap B$ (e.g., Linda is currently active in the feminist movement and a bank teller). The conjunction fallacy corresponds to the case of $F \cap B$ being judged more likely than option B. The disjunction $F \cup B$ is also likely to evaluated by participants as less likely than the individual event F (the "disjunction fallacy"[9]). The conjunction fallacy is also obtained using betting procedures where probabilities are not asked.[10] This may imply a negative interference effect. Here again failure of laws of classical probability necessitates the development of a more general method that can explain this interference effect.

- Overextension of Category membership: Assignment of membership weights on pair of concepts and their conjunction produces what is called the 'guppy effect', the corresponding problem being referred to as the 'pet-fish problem'. The example usually given is that people mark "cuckoo" (when given as an item) as a better representative member of the conjunction 'Bird and Pet' than of the concept of Pet on its own. If the conjunction of logical propositions governs the effect because of the conjunction of concepts, the second should be at least as great as the first, which is not so in practice. This deviation referred to as 'overextension'[11] may also be viewed as interference effect. Define $P(A|x)$ as the probability that category A is true given the item x, and $P(A \cap B|x)$ as the probability that category $A \cap B$ is true given the item x. Then law of total probability suggests that $p(A \cap B|x) < p_T(A|x)$, where p_T is the total probability of A given the item x. But the judgments imply a negative interference effect. Several other examples reveal that this effect is really in abundance. Thus a more holistic approach is needed to explain the overextension of category membership, which occurs due to conjunction of more than one concept.

Memory recognition over-distribution effect and fallacies: A paradigm called the conjoint-recognition paradigm is used for this kind of test. Test participants are first rehearsed on a set T of memory targets (e.g., each being a short description of an event). A recognition test phase follows subsequent to a brief delay after the rehearsal phase. This involves a series of test probes consisting of trained targets from T, related non-targets from a different set R of distracting events (e.g., a set of new events with some relation to a target event), and unrelated set U of non-target items (e.g., set whose members are completely unrelated to the targets). Verbatim (V), and Gist (G) instructions requiring acceptance of only exact targets, or non-targets respectively, are used. Now if a test probe belonging to a target set T is used in the memory test: when a verbatim question is asked, the probability of accepting the target is defined by the conditional probability $P(V|T)$ and for the gist question being asked, the probability of accepting the target is $P(G|T)$. Correspondingly, for a verbatim-or-gist question, we have $P(VorG|T)$. Now a probe x comes from T or G but not both – that is $P(VorG|T) = P(V|T) + P(G|T)$. The difference, $EOD(T) \equiv P(V|T) + P(G|T) - P(VorG|T)$ is an episodic over distribution (EOD) effect. A positive EOD effect was reported from 116 different experimental conditions. 90% of these 116 studies produced this effect, with a mean value of 0.8 for the EOD. Busemeyer *et al.*[4] claimed that quantum probability theory provides a simple and coherent explanation for all six of the above empirical findings of interference effects. Actually, the work of Aerts etc.[12] pioneered the application of quantum theory to cognition and decision. Several authors are trying to explain these findings using Dempster-Shafer theory as an alternate to classical probability theory. It leads us to reanalyze the whole scenario and the applicability of probability theory (both classical and quantum probability) as related to Bayesian belief in contrast to Dempster-Shafer belief. The issue is very important in the cognitive domain in the sense whether the process of perception is inseparable from rational (broadly Bayesian) processes of belief fixation or not and context effects are felt somehow at the intermediate level of processing.

3. Quantum Ontology and Context Dependence

Quantum Mechanics is usually considered as an empirical structure that corresponds to a genuine law of nature, also the laws of quantum logic, which finally lead to quantum mechanics must contain empirical components that at the end of this way of reasoning imply the empirical components in quantum mechanics. Where do these empirical elements in quantum mechanics come from? In the above type of description, the issue is what is meant by the unity of complementary aspects at the cellular level. Can we think of any natural constant which gives rise to this kind of unit (like Planck constant) at the cellular level? Essentially it is needed to construct a framework describing cellular basis of cognition and its relevance to the shift of paradigm.[13] Here interaction of neuroscientists, physicists and mathematicians are needed to build up the required framework. This will shed new light in robotics and artificial intelligence research.

4. Concluding Remarks

The new empirical findings in the cognitive domain as classified into six broad groups clearly indicate the inadequacy of classical probability theory for modeling decision making process. Quantum paradigm seems to be more appropriate for modeling in cognitive domain.

References

1. Jerome R. Busemeyer and Jennifer S. Trueblood: Theoretical and empirical reasons for considering the application of quantum probability theory to human cognition; This material is based upon work supported by US National Science Foundation Grant No. 0817965.
2. Tversky, A. and Shafir, E.: The disjunction effect in choice under uncertainty. *Psychological Science*, **3**, 305–309 (1992).
3. Savage, L. J.: (1954). *The foundations of statistics.* John Wiley and Sons, New York (1954).
4. Busemeyer, J. R., Pothos, E. M., Franco, R. and Trueblood, J. S.: A quantum theoretical explanation for probability judgment errors. *Psychological Review*, **118** (2), 193–218 (2011).
5. Townsend, J. T., Silva, K. M., Spencer-Smith, J. and Wenger, M.: Exploring the relations between categorization and decision making with regard to realistic face stimuli. *Pragmatics and Cognition*, **8**, 83–105 (2000).
6. Busemeyer, J. R., Wang, Z. and Lambert-Mogiliansky, A.: Comparison of markov and quantum models of decision making. *Journal of Mathematical Psychology*, **53**(5), 423–433(2009).
7. Conte, E., Khrennikov, A. Y., Todarello, O., Federici, A., Mendolicchio, L. and Zbilut, J. P.: Mental states follow quantum mechanics during perception and cognition of ambiguous .gures. *Open Systems and Information Dynamics*, **16**, 1–17 (2009).

8. Tversky, A. and Kahneman, D.: Extensional versus intuitive reasoning: The conjuctive fallacy in probability judgment. *Psychological Review*, **90**, 293–315 (1983).
9. Carlson, B.W. and Yates, J. F.: Disjunction errors in qualitative likelihood judgment. *Organizational Behavior and Human Decision Processes*, **44**, 368–379 (1989).
10. Sides, A., Osherson, D., Bonini, N. and Viale, R.: On the reality of the conjunction fallacy. *Memory and Cognition*, 30, 191–198 (2002).
11. Hampton J.A.: unitary model for concept typicality and class inclusion. *Journal of Experimental Psychology: Learning Memory and Cognition* , **14**, 12–32 (1988).
12. Aerts, D.: Quantum structure in cognition. *Journal of Mathematical Psychology*, **53**(5), 314–348 (2009).
13. Sisir Roy, *Decision Making and Modeling in Cognitive Science*, Springer, (2014).

Chapter 11

Knowledge, its hierarchy and its direction

Apoorva Patel

Centre for High Energy Physics
Indian Institute of Science, Bangalore 560012, India
adpatel@cts.iisc.ernet.in

"The purpose of life is to obtain knowledge, use it to live as satisfactorily as possible, and pass it on with improvements and modifications to the next generation." The interpretation of these words may be subjective, yet this is what all living organisms—from bacteria to human beings—do in their life time. Evolution points out the direction in biological systems for acquiring, processing and communicating information. Comparison with the computers we design illustrates the hierarchical architecture of information processing. We are now in a position to make the underlying principles mathematically precise, and to exploit them as far as the laws of physics permit. That will decide which questions are relevant to the future of life and which are not.

1. Introduction

The topic of this workshop, concerning cognition and consciousness, is such that it is hard to avoid philosophical overtones, and I shall make no attempt to do that. But underneath all that I am going to say, I am a physicist, and that will show through in my presentation.

I am going to talk about knowledge—its structure, manifestations and implications, as we have interpreted them over the ages. Let me begin by showing you an instructive example, which is a video of a white blood cell (neutrophil) chasing a bacterium.[1] This process of phagocytosis progresses from physical sensation to decision to action. Both the neutrophil and the bacterium could not have done what they did without a certain sense of awareness of their surroundings and who they themselves are. If you watch closely, you will even observe an instance when the neutrophil makes a choice between which of the two detected bacteria to go after.

Table 1. Hierarchical levels of experience, their level of operation, and the questions they address, according to traditional Indian philosophy.

	Mode of Experience	Level of Operation	Question to be Answered
1	Bodily sensation	Body (शरीर)	What is happening?
2	Sensory perception	Senses (इन्द्रिय)	What is this?
3	Perceptual conception	Outer mind (मनस्)	How come this?
4	Conceptual reasoning	Intellect (बुद्धि)	Why this and not that?
5	Reasoned judgement	Inner mind (चित्त)	What is its meaning or purpose?
6	Judged action	The ego (अहंकार)	What ought to be done?
7	Acted realisation	The self (आत्मन्)	Who am I?

Now consider the hierarchical levels of experience analysed in the traditional Indian philosophy, and summarised in Table 1. The levels progress from inductive to deductive to abductive logic, and neutral to objective to subjective view of reality. With a little bit of thought, we can surmise that phagocytosis exhibits all these seven stages of experience, and the stages are not independent but closely tied together. Acquired knowledge, ranging from immediate sensation to the memory based on prior experiences (going all the way back to what is written in the genes), is at the heart of this process.

To what extent can we understand the interlinking of the hierarchical structure, in this simple example? That will be at the root of how we interpret knowledge and consciousness.

2. The Meaning of Life

What is *the answer* to the ultimate question of life, the universe and everything?

42

according to the Hitchhiker's Guide to the Galaxy.[2] That is anticlimactic, and was meant to be a parody, but is there really a better answer to such a question? All our philosophical enquiries frequently come down to our view of ourselves: *Why are we here? What is so special about us? What is our future?* And so on. Science does have something to say about such an anthropocentric outlook!

Let us look at our location in space. The universe has billions of galaxies, each with billions of stars and a similar number of planets orbiting the stars in the habitable zone. Our sun is an average star occupying a non-descript place in our galaxy (called the Milky Way or आकाशगंगा). On the scale of a printed picture page of our galaxy, the sun would be smaller than an atom. Clearly there is nothing special about the position of our earth in the universe.

Next look at our location in time. The universe is approximately 13.7 billion years old. In the beginning, it was an extremely tiny, dense and hot ball of elementary particles (no atoms). Atoms formed as the universe expanded and cooled. Our sun is about 4.5 billion years old; it is not a first generation star. Life on earth appeared around 3.8 billion years ago. Human beings (Homo sapiens) appeared on earth around 100,000 years ago. If the age of the universe is scaled to a day, human existence on earth would be for less than a second. It is hard to find anything special about the time of our appearance on the earth either.

Let us broaden our view-point and look life itself as a whole. The universe has billions of planets that can support life. Single celled life forms outnumber and outweigh(!) multi-cellular ones, and are ubiquitous. For complex living organisms on earth, average lifetime of a species is about 4 million years. Most living cells seldom last more than a month or so. Individual components of cells are constantly renewed. There isn't a single bit of any of us (not even a molecule) that was a part of us nine years ago! The atoms in each DNA strand get knocked off and replaced, in a continuous jostling with other molecules, ten thousand times a day.[3] There is hardly anything of permanence even for all of life.

So what meaning can be extracted from all this? Atoms are fantastically indestructible as far as life is concerned; they just get rearranged in different ways. Each of us would have a billion atoms that once belonged to the Buddha, or Genghis Khan, or Isaac Newton—may be an exciting or may be a sobering realisation! It is not the atoms themselves but their arrangement, which carries biochemical information. Life is fundamentally a non-equilibrium process. Living organisms evolve, even as the atoms keep on shuffling. To put it in the language of computer science:

Hardware is recycled, while software is improved!

It is the accumulated software, stored in various types of memories, that provides the identity to an individual.

3. Evolution: Direction vs. Goal

As keenly observed by Theodosius Dobzhansky, *"Nothing in biology makes sense except in the light of evolution"*.[4] Over billions of years, biological evolution has experimented with a wide range of physical systems for acquiring, processing, selecting and communicating information. In this history, evolution has been dominated by changes at the genotype level, and selection at the phenotype level. We carry two lasting signatures of this evolution: (a) all living organisms begin life as a single cell, and (b) the genome is inherited as a read-only-memory.

In Table 2, I have described the development of systems for conveying information as living organisms evolved. The process has not stopped, and its continuation hardly requires any persuasion. Look at the number of gadgets that we were not born with but we have become accustomed to carry: spectacles, pens, watches, mobile phones. It is easily noticed that evolution has progressively discovered more and more sophisticated mechanisms, which expand our range of knowledge acquisition and communication. In this natural selection,

- Communication range expands in space and time.
- Physical contact reduces.
- Abstraction increases and succinct languages arise.
- Complex translation machinery develops.
- Cooperation gradually replaces competition.

Table 2.   Evolution of communication systems in living organisms.

Organisms	Messages	Physical Means
Single cell	Molecular (Genes, Proteins)	Chemical bonds, Diffusion
Multicellular	Electrochemical (Nervous system)	Convection, Conduction
Families, Societies	Imitation, Teaching, Languages	Light, Sound
Humans	Books, Computers, Telecommunication	Storage devices, Electromagnetic waves
Gizmos or Cyborgs ?	Databases	Merger of brain and computer

There is a lesson in this pattern. Environmental hazards exist at all scales, and knowledge helps overcome these hazards. As a result, the increasing reach of knowledge has become the driving force behind "survival of the fittest". It provides the direction for evolution, even when the goal is unclear.

Indeed, direction is needed to go forward, but goal is not! The bacteria existing billions of years ago had no clue whether they will evolve into tall trees, fearsome dinosaurs or smart human beings, and we can only guess what living organisms may turn into in future. Evolution has followed a bottom-up approach. That can be fully self-contained, as long as direction is available at every step.

On the other hand, many of our queries arise from a top-down approach. A typical situation would involve finding the direction for a specific goal, which is a non-trivial exercise. Is that the correct strategy for deciding our actions, or a waste of effort and cause for misery? (Note that the dynamics is local in physics, and the space-time range of exploration is always finite, while global constraints provide conservation laws.) The fact that we are better off concentrating on the direction of action, without worrying much about the final result, is forcefully expressed in the following ancient wisdom:

कर्मण्येवाधिकारस्ते मा फलेषु कदाचन
—श्रीमद् भगवद्गीता २ : ४७

Thy right is to work only, but never to its fruits.

4. Living Organisms vs. Computers

With increasing reach of knowledge, human beings have become capable of asking (not necessarily answering) more and more elaborate questions. To put that in perspective, let us compare the hierarchical processing of information in living organisms and in computers. We have a good understanding of what goes on inside computers because we know the basic principles that we have used to design them.

We use the terminology where data represent a particular realisation of the physical system among its many possible states, information is the fungible abstract mathematical property obtained by detaching all physical attributes of the data, and knowledge is the practical outcome obtained by adding appropriate interpretation to the abstract information.

Table 3. Comparison of hierarchical information processing in computers and in living organisms.

↓ Computers		Living organisms ↑
Data	Input	Environmental signals
Pre-processor	High level	Sense organs
Compiler	⇑	Nervous system
Assembler	Translation	Brain
Machine code	⇓	Electrochemical signals
Electrical signals	Low level	Proteins
CPU and memory	Execution	Genome

Table 3 illustrates how in the hierarchical structure of information processing, translation machinery (e.g. compilers) maps high level instructions to low level executable tasks. Subjective, varied and abstract higher levels are irrevocably tied to objective, limited and physical lower levels. Both top-down (↓) or bottom-up (↑) designs, indicating where the fundamental programme is written, are possible. Obviously, genetic approach is bottom-up, while conscious effort is top-down.

Bottom-up construction can make programmable devices, e.g. the fertilized egg knows how to produce a brain without knowing what will be stored in it. Top-down feedback can select and modify the rules, e.g. we can select stem cells and alter the genes towards a specific goal. But neither can work without an understanding of all the intervening levels. As far as life itself is concerned, there is little doubt about what came first; "how to learn" has always been more more important than "what to learn". But we can now look at alternatives.

5. Artificial Intelligence vs. Life

Evolution works by generating a variety of possibilities, and then selecting appropriate ones from them. They do not have to be restricted respectively to genotype and phenotype stages only. In programmable devices, the corresponding features are "imagination (↑)" and "feedback (↓)". (Their successful combination allowed Deep Blue to beat Garry Kasparov in chess.) Let us consider what they involve.

In imaginative exploration, memory (pattern recognition based on past history) has a crucial role. In generation of new possibilities, random choices are convenient at small scale, while mix-up of established features is efficient at large scale. Selective feedback reduces choices and focuses progress. Clever amplification/suppression can specify priorities, deciding what to retain and what to forget. Clarity is increased by digital punctuation of analogue processes.

We are now using these ideas to build intelligent devices. But let us turn the attention back to ourselves. In our experience, children are good at imagination. At that stage, the brain is more programmable, although the failure rate is higher. On the other hand, adults are good at feedback. The brain becomes more filled with memory, and that offers higher security in selecting tasks. Both imagination and feedback are essential for progress. But a balance has to be struck between the two, because too much exploration is wasteful and excessive feedback hinders exploration. When the computers get too cluttered, we save the important stuff, then wipe the slate clean and start all over again. Life's answer to the same conundrum is similar: *The cycle of life and death "reboots" the system.* Contributions of children and adults then alternate.

It is worth noting that every level of knowledge in the hierarchy (see Table 2) has a role to play, and it can be tinkered with using appropriate methods. The changes can be made permanent, provided one can control the translation machinery between the levels. Development of higher levels of knowledge communication, not just the genetic one, illustrates that life has grasped this fact. It can be observed that the capability to pass on knowledge influences life expectancy: For primitive living organisms, reproduction is usually the last stage of life, some advanced ones live longer to take care of their children, and a few (like us) live even longer to look after their grandchildren.

Our efforts to build increasingly powerful and versatile computers have taught us a lot about acquisition, processing and communication of knowledge. It is no longer an exercise of trial and error, but it is a problem of systematic design. It is not unrealistic to say that we are now in a position to figure out the whole machinery of life, and to exploit it as far as the laws of physics permit.

6. Conclusion

According to Gödel's theorem, any consistent axiomatic system allowing recursion cannot be complete, i.e. it will contain precise but unprovable statements. The well-known example is the "halting problem", where a universal computer is unable to determine whether a given programme will halt or loop indefinitely. Life is complex enough to make certain questions about goals unanswerable. Fortunately, questions about directions can always be answered, and that is sufficient for progress. Indeed, all the instructions contained in the genetic machinery are about directions and not about goals.

Thus life's exploration of knowledge will continue, with the hierarchical structure and memory playing indispensable roles, and pointless questions getting discarded along the way. Let me then end with an invocation for peace (शान्तिमन्त्र):

$$...तमसो\ मा\ ज्योतिर्गमय...$$
Lead me from darkness to light.

References

1. This video is taken from a 16-mm movie made in the 1950's by the late David Rogers at Vanderbilt University. It is available at http://www.biochemweb.org/neutrophil.shtml.
2. D. Adams, *The Ultimate Hitchhiker's Guide to the Galaxy* (Del Rey, New York, 2002). Also have a look at today's Google doodle, commemorating Douglas Adams' 61st birthday!
3. Many such interesting observations can be found in: B. Bryson, *A Short History of Nearly Everything* (Black Swan, London, 2004).
4. T. Dobzhansky, American Biology Teacher **35**, 125 (1973).

Chapter 12

Some Remarks on Numbers and their Cognition

P.P. Divakaran

*National Centre for Biological Sciences — Tata Institute of
Fundamental Research,
GKVK, Bellary Road, Bangalore 560065
divakaran@ncbs.res.in*

Humans perceive numbers (positive integers) through counting, i.e., by
establishing a correspondence between the cardinality of a set of objects
and a structured set of number symbols or number names. The structure
requires the choice of a base — 'succession' structure alone is not enough.
The point is made that for numbers below the base, cognition involves
the learning and recalling of this association, fixed once and for all, while
for larger numbers it results from learning the rules of construction of
based numbers. Experiments on number cognition which do not take
into account this essential distinction are likely to remain inconclusive.

First some generalities. Numeracy is such a common skill, almost an element of basic literacy, that we hardly stop to wonder how we actually
perceive a number. ('Number' in these remarks generally means a positive
integer). Abstractly, it is the cardinality of a discrete, finite set. In concrete terms, everyone knows that the way to determine it for any set is to
count the elements in the set, i.e., by setting up a one-to-one correspondence between the objects in the set and a set of *ordered* number symbols or
number names. Obviously, for this theoretical procedure to be of practical
value, we must impose a structure on the set of number representatives,
whether symbols (perceived visually) or names (perceived aurally), so that
the order becomes internalised or 'intuitive'; making an undifferentiated set
correspond to another undifferentiated set is of no help in the cognition of
the precise numerical significance of the cardinality of a set.

Such a structure is provided by a *base* like the universally used 10. A base is a unit or a standard number in terms of which to measure an otherwise 'unknown' number, just as a length is measured in terms of a standard length such as the meter; in fact, counting is the most primitive measurement there is. The choice of 10 as a universal unit or standard is just as arbitrary, in principle, as that of the meter. In practice however, as we shall see later, a base has to meet certain requirements determined by cognitive needs if it is to serve its purpose of facilitating the precise apprehension of large numbers in a sensible way.

The great preoccupation of European mathematics in the late 19th century with questions regarding its logical foundations encompassed naturally an axiomatisation (the Peano axioms) of the notion of numbers as members of a set $\mathbf{N}$. The essence of the axiomatic description of $\mathbf{N}$ lies in formalising the idea of succession: to every member of $\mathbf{N}$ is associated another member, its successor, and there is a unique member, the initial number '1', which is not the successor of any member. The Peano axioms represent the ultimate abstraction, the final logical outcome of a process whose roots lie in the empirical perception of discrete quantity, in other words in a cognitively defined need[a]. But by itself the axiomatisation is of no direct practical use. The axioms do not provide any help in doing arithmetic with concretely specified numbers or even in enumeration which is the essential prerequisite for arithmetic. It is in bridging this gap that a base like 10 becomes indispensable. To 'measure' a number N, we first take away as many (say N_1) multiples of 10 from it as possible, leaving behind a remainder $n_0 < 10$: $N = N_1 \times 10 + n_0$. If $N_1 < 10$, the process ends; otherwise repeat it as many (k) times as necessary until $N_k = n_k$ becomes less than 10:

$$N = n_k \times 10^k + n_{k-1} \times 10^{k-1} + \cdots + n_1 \times 10 + n_0.$$

The ordered sequence $(n_k, \cdots, n_1, n_0)$, without the commas and the bracket, is how we generally represent N in writing. What the choice of a base does is thus to allow every number to be presented as the value of a polynomial when the variable is fixed at 10 and with coefficients that are less than 10. The problem of counting is thereby reduced to the cognition of the 'atomic' (see the next paragraph) numbers 1 to 9 (more generally, numbers less than the base) and, for numbers larger than 10, the

---

[a]Some of the recent theoretical work on number perception by cognitive scientists has the uncanny air of being paraphrases, in a logically less precise language, of the Peano axiomatic framework. The line of thought can be traced back from, for instance.[1] In my view this is putting the cart before the horse.

application of an algorithm involving rules for multiplying powers of 10 by the atomic numbers and adding up the results.

Decimal enumeration, the use of 10 as the base, originated in India, as is well known. It is less well known that this happened in pre-Vedic times, before ca. 1,100 BC, as attested by the grammatical rules employed to impement the operations of multiplication and addition in the formation of number names in the *Ṛgveda*. The analysis of these number names turns out to be a very rewarding exercise and has had an impact on several aspects of Indian epistemology over a long period. In particular, the special role played by the numbers up to 9 in number nomenclature was recognised, most notably by the linguist-philosopher Bhartṛhari (5th–6th century AD), who likened the coming together of such numerals to form larger compound numbers to the gathering of atoms to make gross matter (as well as to the construction of sentences from 'atomic' words).[b]

Fixing a base makes rule-bound arithmetic possible. To appreciate what this means, it is helpful to remember that we can know the results of adding and multiplying atomic numbers only by experimentation, real or in the mind, followed by an act of memorisation (think of the multiplication tables of our childhood). Once that is done, arithmetical operations on large numbers (long multiplication for example) are carried out using rules, few in number, which are in turn learned and memorised. These general rules depend implicitly on the polynomial representation as they are essentially the rules for adding and multiplying polynomials. Historically of course people did arithmetic for millennia before its structural foundation in the algebra of polynomials, and indeed the notion of a polynomial itself, was identified. (Both in India and in Europe polynomials and related algebraic objects were first introduced in analogy with decimal numbers). The fact remains that the structure underlying the manipulation of decimal numbers and polynomials is the same, as was acknowledged in India already in the 16th century and, in an insightful passage in his work with infinite series, by Newton. The important message is that, while arithmetical skill is empirically acquired as long as it has to do only with atomic numbers, it becomes rule-based and algorithmic when dealing with numbers larger than the base.

But what concerns me more here are the implications of the polynomial or place-value representation for the cognition of numbers themselves, before we do any arithmetic with them. It is completely obvious that counting

---

[b]For a reconstruction of the state of numeracy in Vedic India and related issues, see Ref. 2.

up to the base and counting beyond the base call upon very different mental processes.[c] To anchor the discussion to our daily practice, let us assume that we count by writing down the ordered string of symbols for the atomic numerals that represents a number. With enough exposure and familiarity, our sense of the number represented by the string 123 for example might appear to be instantaneous. It would seem reasonable to suppose, nevertheless, that the recognition is a result of learning, and recalling at will, that it really stands for the result of running the 'place-value algorithm': $123 = 1 \times 10^2 + 2 \times 10 + 3$; the mental faculties involved are those of reasoning and memory, of being able to apply precise (and well-remembered) rules, without variation, time after time. For the atomic numerals, in contrast, there can be no rules to learn. The symbols themselves can be chosen entirely arbtrarily: any nine distinct symbols and any ordering among them (together with a symbol for 0 in written numbers) will do as long as the symbols and their ordering are the same for everyone[d]; after that it is all down to memory.

These considerations hold just as well for realisations other than the symbolic or written (which is then read), for instance the oral realisation in which the atomic numbers are given (again, completely arbitrary) names that are spoken (and heard), once again in an unvarying order – whether the names are also written down makes no difference. To form the names of large (composite) numbers, they are then to be combined grammatically in a way respecting the arithmetical processes at work. That is how decimal number names in the *Rgveda* are constructed, as I mentioned earlier. Indeed it is useful to think of a fundamental *place-value principle* as existing in the abstract, of which the symbolic-written and the nominal-oral (and conceivably others) are representations.

The point of these remarks is that it is this abstract concept of based numbers, irrespective of how different societies may have chosen to express

---

[c]I should stress that by counting I mean the exact determination of cardinality, not some approximate estimation of rough magnitudes as in several anthropological studies. It is not obvious to me that the two accomplishments need have much in common other than that the behavioural responses through which we assess their reliability use the same means, namely a knowledge of number symbols or number names.

[d]It is an interesting fact that in early societies with writing (Old Babylonia, Indus Valley), the symbols for atomic numbers consisted of the appropriate number of repetitions of a token representing 1: the familiar 'cocktail glass' in Babylonia and a vertical stroke in the Indus Valley. Replacing such literal number symbols by more abstract ones (4 instead of |||| for instance) marks a major evolutionary step towards greater abstraction. It is nevertheless amusing that the modern 'Arabic' numerals 1, 2 and 3 are cursive variants of the corresponding number of (vertical or horizontal) strokes.

it, that I feel should be the primary object of the cognitive study of our number skills. My reason is that, simply by virtue of being abstract, it is universal and free of the cultural context, the very qualities that, we would expect, characterise a faculty that is innate. There has of course been an enormous amount of work done in this field recently,[e] spanning a very wide range of approaches, theoretical, observational and experimental – both behavioural and neuro-scientific – but there remains a strong case for refining them to the point where distorting or irrelevant factors, for example the various ways numerical information is presented to the subject in an experiment, are minimised; in other words, in a manner that facilitates the understanding of how numerical notions are internalised in as abstract a way as possible. An extreme case of what to avoid would be the use of conventional number symbols as that depends on prior associations and really only checks the role of memory in recognising 'numerosity' (a word popular with cognitive scientists studying numerocity, not always spelt the same way).

The idea that the counting of small numbers (technically, those below the base) is a matter of intuition, an innate faculty, has a very long history. The Sanskrit noun for number, *saṃkhyā*, comes from attaching the prefix *sam* to the verb root *khyā* which in the Vedic literature signifies 'to see'; so *saṃkhyā* literally means 'that which is seen together', taken in at a glance as it were. Along with over two thousand occurrences of number names, the *Ṛgveda* also has many passages which attribute the power to see objects "minutely and precisely" to Agni, the fire god who shines and illumines; it is as though there was no other way to discriminate among numbers except through the innate gift of clear vision. (Interestingly, there is a similar linking of fire and numeracy in the Greek myth of Prometheus).

The distinct mental processes that we bring to bear on the perception of numbers smaller and bigger than the base should clearly have an influence on the determination of the ideal base, since the former set can only be 'intuited' while the latter can be computed. The adult human ability to 'take in' accurately (without conscious counting) the cardinality of a (say, visually presented) set of objects is limited to relatively small numbers, as we know from daily experience. There is a fair amount of not very rigorous but persuasive evidence that this number is about 8 to 10, at least among those who have a sense of numbers that are not too small, perhaps a few more if we impose some regular pattern on the set. (I might add that the

---

[e]Two reviews out of very many which may be consulted for further information are Refs. 3 and 4.

Indus Valley culture, very likely, counted to the base 8). It will be nice if such anecdotal theorising about the ideal base can be replaced by reliable and reproducible experiments aimed at distinguishing between 'intuiting' and counting. In a certain sense, what decimal place-value counting does is to impose an infinitely recursive abstract pattern on the set of all numbers (up to 'infinity') and this it achieves by using the procedure of counting by tens repeatedly.

The number names of the *Ṛgveda* hold another point of interest from the cognitive perspective, which is that they are names (it is widely acknowledged that the Vedic culture was oral) and therefore conditioned by the language in which they are expressed and combined. It is in fact impossible to study the development of a number sense among the early Vedic people without getting heavily involved in subtle questions of language and grammar. There is a school of thought that holds to the view, generally based on field work in orally literate primitive societies, that the human cognition of numbers is inextricably tied to language skills but there are also those who question such a connection. Similarly, linkages are sometimes sought to be established between our sense of numbers and the sense of space (in a plane or between right and left). For those belonging to a dominantly 'written' culture, it will be surprising if such correlations did not exist. Once again, the case for making sure that experiments on the human cognition of numbers are kept free from cultural noise and focussed on the abstract mathematical ideas involved is a strong one.

Acknowledgment

I give warm thanks to Vidyanand Nanjudiah for his comments on a draft of this note.

References

1. A.M. Leslie, R. Gelman and C. R. Gallistel, "The Generative basis of natural number concepts", Trends in Cognitive Sciences **12**, 213-218 (2008).
2. B. Bavare and P. P. Divakaran, "Genesis and Early Evolution of Decimal Counting: Evidence from Number Names in Ṛgveda", Ind. J. History of Science (to appear).
3. S. Dehaene, "Origins of Mathematical Intuitions: the Case of Arithmetic", Ann. N. Y. Acad. Sci. **1156**, 232-259 (2009).
4. L. Malafouris, "Grasping the concept of number: How did the sapient mind move beyond approximation?" in I. Morley and C. Renfrew (eds.), *The Archaeology of Measurement*, Cambridge University Press (2010).

Chapter 13

Conceptual Revolution of the 20th Century Leading to One Grand Unified Concept — The Quantum Vacuum

B.V. Sreekantan

National Institute of Advanced Studies,
Indian Institute of Science Campus, Bangalore 560012, India
bvsreekantan@gmail.com

"The Mathematics of Riemann's geometry aligns perfectly with
the physics of gravity." – Albert Einstein

Concepts and the relations between concepts are the basis for all our scientific understanding and explanation of the wide variety of constituents and phenomena in nature. Some of the fundamental concepts like space, time, matter, radiation, causality, etc. had remained unchanged for almost four hundred years from the time of the dawn of science. However all these underwent a drastic transformation in the 20th century because of two reasons. One, in the light of certain experimental findings two radical theories namely theory of relativity and theory of quantum mechanics replaced the classical theory that had dominated since Newton's time. Secondly, the science-technology spiral resulted in the discovery of very many new features of the universe both on the micro scale and on the mega scale. There was an exponential increase in our knowledge. These new facts could not be fitted into the old concepts. Apart from drastic revision, many new concepts had to be brought in. Despite all this, one very encouraging trend has been to discern a holistic synthesis and unification of the different concepts — an endeavor that has been helped by experiments over a wide scale of energy and distances and most importantly from theoretical insights triggered by mathematical underpinnings. These developments in physics and astrophysics are pointing to one grand concept, namely, the " quantum vacuum" endowed with certain special properties, as the substratum from which all the constituents of the universe as well as the processes of the universe emerge, including the creation of the universe itself. This is the view, at least of some of the scientists. In this brief article the essence of these approaches

toward unification is highlighted. Maybe life sciences can take a clue from
these developments in physical sciences.

1. Introduction

Be it on the basis of common sense or on the basis of most rigorous science,
the understanding and explanation of all phenomena that we experience
near us, far away from us or inside us, depend on a variety of concepts and
the relations between the concepts.

Some of the oldest concepts which have been in use in normal parlance
of people and for which one has evidence even from the scriptures that date
back to thousands of years, are space, time, matter, mass, heat, light, force,
energy, vacuum, causality, etc.. They continue to keep the same meaning
in their usage in daily life even today. However, these concepts also became
part of science about four hundred years ago during the time of Bacon,
Descartes, Newton, Galileo — the founders of modern science. Concepts
were gradually quantified; units introduced to enable meaningful measurements of high accuracy with instruments developed for the specific purposes. New concepts like velocity, acceleration, momentum, kinetic energy,
potential energy etc. were also introduced. Most importantly this helped
in mathematical formulation of the results of observation and prediction of
new results, embodying operating principles of nature figured out through
experience, sometimes intuition. In a sense, the scientific methodology got
initiated at this stage.

Introduction of more and more sophisticated instruments, leading to
systematic studies on the properties of matter, heat, light, sound, etc.,
resulted in increased knowledge, for the explanation of which, many new
concepts like molecules, atoms, entropy, refraction, diffraction, interference,
spectral lines, acoustic waves, light corpuscles, light waves and so on had to
be postulated. These concepts dominated the fields of Mechanics, Heat and
Thermodynamics, Optics and Acoustics. These became the most prominent subjects of experimental and theoretical studies till the end of the
19th century.

Yet another field that came into prominence during this period was the
subject of electricity and magnetism following the classic experiments of
Faraday, Oersted, Ampere, Coulomb and others. A grand synthesis of all

this work in the field of electricity and magnetism was achieved by Maxwell and led him to the recognition of the existence of electromagnetic fields in space, strangely moving with the same velocity as that of light. This fortuitous coincidence with the velocity of light led to the identification of light as an electromagnetic (EM) wave. Subsequently Hertz produced EM waves of longer wavelength in the laboratory.

The concept of field (magnetic, electric and electromagnetic field) introduced in this context, presented right from the beginning, a serious ontological problem. A wave amplitude could be a mathematical entity but since the EM wave also transmitted energy it had to be a physical entity. Maxwell tried to identify the EM waves as 'ether waves', but did not succeed. We shall come back to the question of 'field' again later.

Towards the final years of the 19th century a new concept "particle" as a fundamental entity burst forth into prominence and has reigned supreme in the field of physics since then. The role of particles at the fundamental level became evident through the detection first of the emission of the "alpha-particle" by uranium nucleus, and later detection of other particles — X-rays, electrons, and protons in the study of phenomena associated with discharge tubes. The electron turned out to be the particle responsible for electric current, the proton the particle in the nucleus of atoms and X-rays — the particle of electromagnetic radiation called the 'photon'. The EM wave/particle (photon) signalled the beginning of acquaintance with another new feature — the same entity being both a wave and a particle — (a transcendence beyond common sense) — the complementarity principle was advanced by Niels Bohr.

2. Space and Time

Till the advent of Special Relativity in 1905, space and time had been regarded in complete conformity with common sense, as autonomous, independent of the external surroundings and their movements. In fact Newton in his Principia had said: "Absolute Space", in its own nature without relation to anything external always remains similar and immovable and similarly "Absolute Time" in its own nature flows equably without relation to anything external.

This absolute nature of space and time was torpedoed completely by the special and general theories of relativity. Time became a dynamical entity which could stretch or contract and even stop as for example near a black hole or a singularity. The clock rates depended on the state of motion and on the curvature of space in which they are located. This also meant that we had to give up the concept of a "universal present moment". What is the present moment in one location could be past or future in other locations. A meter stick travelling with a high velocity would shrink in its length and the extent of shrinkage depends upon the velocity.

Such changes in the concepts of space and time were necessary to explain the experimental fact that the velocity of light was a constant, independent of the motion of the source or observer. This straightaway contradicted the common sense notion of addition of velocities which had become a law in Newtonian dynamics. According to Einstein the three dimensional Euclidean space and the one dimensional forward moving time had to be fused into a new concept of *space-time* having four dimensions. The geometry of this four dimensional space-time and its special properties are responsible for several features observed in the three dimensional non-Euclidean space and unidirectional time.

Arthur Eddington wrote of this transformation in a transparent and lucid way as long ago as in 1920:[5]

> "The four dimensional world is no mere illustration. It is the real world of physics. An observer on the earth sees and measures an oblong block, an observer in another star contemplating the same objects finds it to be a cube. Shall we say the oblong block is the real thing and the other observer must correct his measures to allow for his motion? All appearances are accounted for if the real object is four dimensional and the observers are measuring different three dimensional appearances or sections and it seems impossible to doubt this true explanation."

3. Matter

Another important concept that has dominated the field of science as well as philosophy is 'Matter', which is the dominant perceptible constituent of nature. The first classification of matter is animate and inanimate, namely, living and non-living. Since the 17th century onwards, chemists have established that all matter on earth consists essentially of 92 elements — hydrogen to uranium, and spectroscopic studies of the stars have revealed that the same elements constitute the stars also. All animate matter too comprises of the same elements; only the molecular structures of the long organic molecules are different.

A very exciting finding of the 20th century was that all elements consist of the same three particles: protons, neutrons and electrons. A further significant revelation was that the protons and neutrons consist of three fundamental particles namely quarks. *Thus today we can say that all matter in the universe consists of quarks and electrons.* Besides the protons, neutrons and electrons, a large number of other short lived particles were discovered in cosmic rays and at high energy accelerators. Among all the particles there are two categories – those that have half integral (1/2, 3/2) spin and those with integral spin (0,1,2) spin. The former are called Fermions since they obey Fermi-Dirac statistics and the latter are called Bosons since they obey Bose-Einstein statistics.

Thus in the matter of matter, a great synthesis, simplification and unification has been achieved in the 20th century. The next question is about the forces that bind the quarks into protons and neutrons, and the protons and neutrons into the various nuclei to form atoms and molecules, and the forces that bind matter to matter to form the universe.

4. Force Fields

It is now well known that four different types of forces operate in nature. They are (1) the gravitational force, (2) the electromagnetic force, (3) the nuclear or strong force and (4) the weak or radioactive force. The concept of gravitational force was introduced by Newton through the famous equation $F_G = G(m_1 \times m_2)/r^2$, where m_1 and m_2 are the masses of any two objects and r is the distance of separation and G is the universal gravitational constant. This is the equation that provided the explanation of the famous Kepler's elliptical orbital motions of planets. The other famous equation of Newton is $F_1 = M \cdot a$ where 'm' is the inertial mass and 'a' is the acceleration.

Newton had no idea how the gravitational force was generated and how it acted at a distance. This 'action at a distance' problem remained unanswered for four centuries till Einstein came on the scene, with his general theory of relativity. According to this theory, the space around any massive object gets warped and the extent of warp depends on the mass of the object and the distance. Using the principle of 'least resistance paths', Einstein was able to replace the idea of gravitational force in Newtonian physics by introducing space and time as dynamic players giving a new mechanism for the operation of gravitational action. This eliminated the action at a distance problem.

According to quantum field theory, every elementary particle – the quarks and anti-quarks, leptons, electrons, muons, neutrinos and τ-meson and their anti particles, are represented by quantum fields which not only give the probability of finding the particle at any given space and time, but also are the repository of other properties like spin, mass, charge of the particles. These properties manifest themselves in interactions with the force fields. The force fields are the electromagnetic field, the gluon field and the gauge boson fields which are the fields corresponding to the photon, gluon, and $W^{\pm}, Z^0$ whose exchange mediate the electromagnetic, strong and weak interactions.

According to the standard model of particle physics, which incorporates all these features, it became necessary to introduce one more field, known as the Higgs Boson field for two specific purposes :

(i) the very short range of the weak interaction ($< 10^{-17}$ m) indicated that the mass of the particles ($W^{\pm}, Z^0$) that mediate the force had to be in the range of 100 GeV. The standard theories, however, indicated that these should have zero mass like the photon. The 100 GeV mass range was confirmed by experiment. To give this mass to the $W^{\pm}, Z^0$, the new field namely Higgs field was introduced as part of the quantum mechanical vacuum field similar to the other fields of the standard model.

(ii) The second purpose which the Higgs boson served was to give masses to all particles which have non-zero mass – all the quarks, leptons etc. The most familiar particle, the electron gets it mass when the electron field interacts with the Higgs field. The mass is not given once and for all. It has to keep on interacting with the Higgs field as long as it exists wherever it be.

The Higgs theory does not give the strength of the coupling constant of interaction. It has to be adjusted with the experimentally determined mass of the electron. The same is the case with all other particles. The theory could also not give the mass of the Higgs particle itself. Since 1970's the Higgs particle has been searched for, at all particle accelerators — Fermilab, CERN etc.. It was clear that the mass exceeded the maximum mass they could produce. On the basis of various considerations the particle physicists came to the conclusion that Higgs mass may be in the range 100–150 proton masses. A new accelerator known as the Large Hadron Collider (LHC) at CERN was built at a cost of about 10 billion dollars through international collaboration.

As of now there are strong indications that at the LHC, a new particle with a mass of 125–126 GeV/c^2 has been discovered and the particle is a boson. While this is strong evidence for the Higgs boson, there is some doubt lurking because of the fact that the ratio of the different decay modes of the Higgs predicted by the theory is not agreeing with the experimentally observed ratio. The statistics is still small. Many believe that the Higgs has been discovered. If so, it is certainly a great triumph for both theory and experiment and for the Standard Model, though the model has still to go a long way to be fully established. Even now values of many parameters in the model have to be put in by hand on the basis of experimental results and are not calculable by theory.

5. 20th Century Astrophysics and Cosmology

The concept of the "universe" itself changed drastically in the 20th century. Even in the year 1915, when Einstein completed his work on the general theory of relativity which had implications on our understanding of the nature of space, time and gravitation, and by which time thousands of stars had been observed with small telescopes, there was still the impression that the universe was static – that the the stars were all in fixed locations, the only movements being those of planets, comets and meteors.

However, the solution to Einstein's first version of his general theory equations showed that the universe was not static – it was either expanding or contracting. To save this embarrassment, Einstein introduced a constant in his equation called the cosmological constant (λ) which Lemaitre interpreted as an "exotic" form of energy that uniformly and homogeneously fills all of space – the energy of space itself, and which has a gravitationally repulsive effect.

It so happened that in 1929 with the aid of the newly installed 100″ telescope at Mount Wilson, Hubble made the astonishing discovery that the galaxies are running away from each other and the speed increases with the increase in the distance between the galaxies. This discovery made Einstein remove the constant λ from his equation. The introduction of λ, he himself called later his 'greatest blunder'. Surprisingly, recent discoveries – that the universe is not only expanding, but also accelerating at a rate higher than what was figured out before, have resurrected "λ" – the cosmological constant.

The discovery that the universe is accelerating was made only a decade ago. This came about by observations of the red shifts of a particular category of supernovae at very large distance from us. The decrease in intensity with distance was more than that expected on the basis of the inverse square law of intensity variation. This acceleration is attributed to the repulsive force of a universal energy which has been called the "Dark Energy" — perhaps the same as what we discussed above as the λ introduced by Einstein, though some differences between λ and this energy are claimed. Dark energy has been identified by some as the universal Higgs field energy which has all the properties required for explaining the acceleration of the expanding universe – repulsive gravitational force and constant energy density in the new areas created by expansion of the universe.

What is most intriguing about this dark energy is that it constitutes 70% of the total energy of the universe with another 25% of "dark matter" energy surrounding galaxies, leaving only 5% as the contribution from the well known baryonic and radiation contents of the universe. The dark matter and the dark energy make their presence felt only through gravitational interaction with known matter and energy. There is no other signature discovered yet.

In his book 'Cosmic Impressions' Walter Thirring describes the drama of creation of the universe as follows:

> "A miniscule speck in an embryonic space field with an Ur-substance of dark energy with negative pressure was the beginning of it all. Repulsion it created swelled this tiny fragment of space, but the Ur-substance regenerated itself everywhere so that more and more of this miracle stuff was formed. The repulsion grows beyond measure and the expansion became an explosion of unimaginable force, the big bang. It shot open

the gates to the underworld over wider and let this matter come bursting forth. This matter developed into the particles we have today, all 10^{88} of them. It had energy and positive pressure so that it took over the reigns from the Ur-substance and slowed down the explosion. However, there was still enough momentum to continue expanding for another 10^{10} years. We are left with a pale reflection of the beauty of the underworld; its symmetry was crushed and destroyed in the big bang. The unified force was split into four forces which continued to be more and more alienated from one another."

6. Higher Dimensional Spatial Geometry & Physical Properties

All our observations and experiments are carried out in three dimensional space and one dimensional forward moving time with which we are all familiar. We have seen that because of the special properties of space and time, Einstein had to introduce the concept of four dimensional space-time and formulated his equations of general relativity in four dimensions.

Kaluza went one step further and showed that Maxwell's equations fall out of Einstein's gravitational equations if he formulated the same in five dimensions. Klein pointed out that this fifth dimension could be an extremely curled-up dimension of the size of Planck length ($\sim 10^{-33}$ cms). Though there were serious problems with the Kaluza-Klein formulations, which were realized later, there were two important new insights that came out of this effort.

(i) The electric and magnetic fields (vibrations of electric charge) observed in the (3+1) dimensional space is due to the geometrical patterns of higher dimensional space. Thus the concept of charge acquired a new angle, similar to curvature of space replacing gravitational force in Einstein's theory.

(ii) The additional dimension could be of miniscule size $\sim$ Planck length, approximately 10^{-33} cms.

Higher dimensional spaces can lead to more symmetries and therefore more conservation laws; and to more physical properties. These ideas have been profitably incorporated in 'string theory' which has six additional dimensions of Planckian size, in addition to the four extended normal four space-time dimensions.

In the multidimensional quantum mechanical theories the spatial patterns (Calabi-Yau spaces) transform as properties like mass, charge, colour charge, spin etc. which were regarded as hidden intrinsic properties in the old theories. *Thus all the properties with which we are familiar in 3+1 dimensional space which have distinctive characters and serve different purposes are just geometrical patterns in higher dimensional space!* Since space is nothing but the collection of many different quantum mechanical waves of energy, in a true sense all the material universe — its constituents and forces are just manifestations of one reality, namely energy in the form of the quantum mechanical vacuum.

7. The Impact of these Concepts on Life Sciences

If as most scientists believe, life and consciousness are also just physico-chemical activities of the bio-molecules and the neuronal chemicals and electrical pulses in the networks of neurons, then the ultimate explanation for these features also will naturally reduce to something similar to what is stated above for the acquisition of physical properties.

However in establishing the relation between the physical properties and quantum mechanical waves in the vacuum, there was considerable help and insight from mathematics which so far is not available for biological processes at the fundamental levels yet. Considering the complicated nature of the issues connected with life and consciousness, it may be necessary to introduce higher dimensions and new wave functions in the vacuum endowed with special properties, that interact in the environments of the cells, cortices, neuronal networks etc. Such an approach will require more subtle level investigations in the bio-sciences both experimentally and theoretically. One has to go far beyond the molecular levels.

A chief objection to the use of quantum mechanical approaches to problems like life and consciousness was that these phenomena require coherence and non-local effects for long periods and at warmer temperatures compared to the nano-scales and ultra low temperature environments in laboratory experiments, in which quantum effects are manifested and tested. However this objection is no longer valid, since evidence is mounting that some of the natural processes like the highly efficient photosynthesis in plants and bacteria turning sunlight, carbon-dioxide and water into organic material, are taking place at normal temperatures.

Another remarkable phenomenon for which there was been no satisfactory explanation i.e., 'homing' of birds migrating over long distances, has recently found explanation in terms of exotic quantum mechanical processes taking place in the bird's eye, again at normal temperatures. Such magnetic field sensing has also been noted in some insects and plants. Philip Ball in his feature article in Nature magazine has called these developments as "the dawn of Quantum Biology".

8. Conclusion

A variety of concepts were introduced in the realm of science for understanding and explaining the structure, contents, and processes in nature over the past several hundred years. As technology advanced, the knowledge gained increased exponentially. It became necessary to introduce many more new concepts. At the same time it also became possible to unearth the intricate inner relations between, what in the first instance, appeared to be very different concepts. This led to unexpected synthesis.

We have discussed in this paper how space, time, matter and energy got synthesized as also the different forces – gravitational, nuclear, electromagnetic and weak. The 20th century physics is strongly pointing towards one ultimate concept namely quantum mechanical vacuum "as the ocean of reality which is composed of waves of energy endowed with specific potential properties and whose spontaneous fluctuations are responsible for all creation and all activity that we see around us".

This does not mean that we have understood everything that needs to be understood. This ocean has the facility to have many more waves that may be needed to explain phenomena not understood yet and those that will be discovered in the future.

References

1. Brian Greene, *The Fabric of Cosmos*, Penguin Books, U.K. (2004).
2. Brian Greene, *The Elegant Universe*, Vintage Books, New York (1994).
3. Ronald Omnes, *Quantum Philosophy*, Princeton University Press, Princeton (1999).
4. Frank Close, *Nothing*, Oxford University Press (2009).
5. A.S. Eddington, *Space, Time and Gravitation: An Outline of the General Theory of Relativity*, Cambridge University Press (1920).
6. Walter Thirring, *Cosmic Impressions*, Templeton Foundation Press, Philadelphia, London (2004).
7. Philip Ball, The Dawn of Quantum Biology, *Nature*, **474**, 272 (2011).

Chapter 14

Classical Coherence, Life and Consciousness

Partha Ghose

*Centre for Astroparticle Physics and Space Science (CAPSS),
Bose Institute, Kolkata 700091*

partha.ghose@gmail.com

There have been many claims that quantum mechanics plays a key role in the origin and/or operation of biological organisms, beyond merely providing the basis for the shapes and sizes of biological molecules and their chemical affinities. These range from Schrödinger's suggestion that quantum fluctuations produce mutations, to Hameroff and Penrose's conjecture that quantum coherence in microtubules is linked to consciousness. I review some of these claims in this paper, and discuss the serious problem of decoherence.

1. Coherence and Decoherence

In physics, coherence is an ideal property of waves or vibrations that enables stationary (i.e. temporally and spatially constant) interference. Coherence is generally possible in physical systems that are linear and allow superpositions, like *waves*, but *not localized objects* like particles.

Waves and their coherence have been the hallmark of quantum mechanics that has made it so different from the Newtonian mechanics of localized material particles with forces at a distance between them. Quantum mechanics has been outstandingly successful in understanding the structure and stability of matter. Yet, the extreme fragility of quantum coherence makes it problematic for applications to open macroscopic systems. The coherence evident in living systems that are open and macroscopic is therefore a theoretical challenge. Nevertheless, there have been serious attempts to understand this biological coherence, and even cognition and consciousness, as essentially quantum phenomena. In this paper I would like to advocate an alternative approach based on classical coherence which is robust.

2. Classical Coherence and Entanglement

Unlike classical Newtonian physics, classical field theory shares some fundamental features like superposition and coherence with quantum mechanics. Interference and diffraction of coherent light have been the main arguments in favour of the classical wave theory of light. It has recently been shown that even *entanglement*, which was considered to be possible only in quantum systems, is inevitable in such theories as well, and it is even possible to carry out many information processing tasks using classical optics that were thought to be possible only in quantum systems. In fact, birefringent crystals have been known since the 17th century to produce states of the form $Au \otimes s + Bv \otimes p$ where u is the path with s polarization and v with p polarization, A and B being amplitudes. In this case, the spatial degree of freedom of a light beam (its path) is 'entangled' with its polarization degree of freedom in the sense that it is not possible to write the 'state' of the light beam as a product of these two degrees of freedom–they lose their identities in this form of light and become non-separable. Recently, the essential role of this non-quantum entanglement in resolving a certain basic issue in classical polarization optics has been pointed out, and nonseparable cylindrically polarized laser beams have been extensively studied and used. The predicted violation of a Bell-like inequality by polarization-path nonseparable/entangled classical states of light has also been experimentally verified. Hence, quantum theory is no longer the only theoretical framework available to study the problem of coherence and cognition in living systems. From a pragmatic point of view, it might, in fact, be more fruitful to view them as emergent phenomena in complex classical systems that show coherence and entanglement.

3. Classical Fields, Oscillators and the Brain

It is well known that classical fields are infinite collections of harmonic oscillators. This can be easily seen as follows. Consider the Hamiltonian of the electromagnetic field,

$$H = \frac{1}{8\pi} \int d^3x [\mathbf{E}^2 + \mathbf{B}^2] \tag{1}$$

where $\mathbf{E}$ and $\mathbf{B}$ are the electric and magnetic fields. Identifying $\mathbf{E} \equiv 4\pi\mathbf{p}$ and $\mathbf{B} \equiv 4\pi\mathbf{q}$, this can be written as

$$H = \int d^3x \frac{1}{2} [p^2 + q^2]. \tag{2}$$

One can discretize it and write it as

$$H = \frac{1}{2} \sum_a^{N=\infty} [p_a^2 + q_a^2], \tag{3}$$

which is an infinite collection of harmonic oscillators of different frequencies.

The human brain is known to be a complex network of a large number (about tens of billions) of interconnected neural oscillators. It is therefore a system that should exhibit 'collective oscillations' of various types depending on the boundary conditions. Each collective mode, called a 'normal mode,' is an independent harmonic oscillation of a characteristic frequency. The most well known of these oscillations are the $\alpha, \beta, \theta, \delta$ brain waves recorded by electroencephalographs. There is also evidence of synchronous oscillations in the cerebral cortex. Such oscillations have been linked to cognitive states, such as awareness and consciousness.

That clearly indicates that it should be possible to model the cognitive states of the human brain by the collective states of an infinite collection of oscillators, i.e. by a *classical field* with a Hilbert space structure, allowing superposition of states and coherence. If that is true, it would point to a new mathematical theory of emotions.

4. Theories of Emotions

Although the well established basic emotions, characterized by universal facial expressions in all cultures, come in three pairs of opposites (Happiness–Sadness, Surprise–Disgust, and Anger–Fear), according to Ekman, all emotions are equally basic. It is also common experience that the human mind is often in a superposition of different emotional states, i.e. in ambiguous states. A fundamental theory of emotions must therefore have a mathematical structure that correctly and completely captures this feature. A 2-dimensional complex Hilbert space is a natural choice for the basis of such a theory. Not only quantum theory, classical field theories like polarization optics also have this structure. The most elegant and complete representation of polarization states in classical optics is the Poincaré sphere, every point of which uniquely represents a polarization state of light, the surface of the sphere specifying all possible polarization states. An analogous unit sphere can be constructed for human emotional states with the three axes X, Y, Z representing three pairs of basic emotions. The surface of the sphere would completely specify all possible emotional states, and the symmetry of the surface would ensure their equivalence. It would also automatically specify the transformations required to take one emotional state to another, which should be of considerable heuristic and therapeutic value.

The fundamental predictions of the theory would be the occurrence of (1) interference effects and (2) contextuality. The experimental tests of these effects in the area of emotions associated with music are being planned in the Sir C. V. Raman Centre for Physics and Music, Jadavpur University.

In addition, to bring out the special characteristics of lyrical music *vis a vis* other forms of emotional stimuli, it would be instructive to measure 'arousal' by four different kinds of stimuli, namely

(1) Prose reading
(2) Poem recitation with the same information content
(3) Songs with the same lyrics
(4) Instrumental music with the same melodic structure as the lyrical songs.

5. Consciousness

Anything that we are aware of at a given moment forms part of our consciousness, making conscious experience at once the most familiar and most mysterious aspect of our lives.

Max Velmans and Susan Schneider

One has to admit that all definitions of consciousness involve circularity or fuzziness, and hence consciousness studies are riddled with ambiguities from the start. In this context, it is amazing to find in the ancient text *Māndukyopanishad*, a part of the *Atharvaveda*, a clear approach, reminiscent of the 20th century school of philosophy known as *phenomenology* founded by the philosopher-mathematician Edmund Husserl (1759-1938). Phenomenology is the philosophical study of the structures of subjective experience and consciousness. The Upanishad starts off by specifying four states of consciousness as the basis of all phenomena:

(1) *Jāgaritasthānah* (A): waking state, *vaisvānarah* (common to all men), *vahihprajnah* (outwardly cognitive, Brentano's and Husserl's 'intentional', i.e. directed at an object, the intentional object), *sthoolabhuk* (enjoyer of gross objects)

(2) *Svapnahsthānah* (U) : dreaming state, *antaprajnah* (inwardly cognitive, so still 'intentional'), *praviviktabhuk* (enjoyer of the mental impressions), *taijasah* (literally, a shining element)

(3) *Sushuptasthānah* (M): deep sleep state, with no desires, *ekeebhutah*
(all experiences unified), *prajnānaghanah* (cognition reduced to an
indefinite mass), *ānandamayah* (full of bliss), *ānandabhuk* (enjoyer
of bliss), *chetomukhah* (gateway to definite cognition), *sarvesvara* (the
originator of all phenomena experienced in the waking and dreaming
states), and

(4) *Turiya*: negation of all, *prapanchopashamam* (where all phenomena
cease), *shāntam* (peaceful), *shivam* (blissful), *advaitam* (non-dual),
sa ātmā (it is the self), *sa vijneya* (it is to be realized).

The letters A (for *akāra*), U (for *ukāra*) and M (for *makāra*) denote
the first, the middle and the last elements of the famous syllable *Om* which,
in its partless aspect, is the fourth state which is transcendental and devoid
of all phenomenal existence. It is neither inwardly cognitive, nor outwardly
cognitive, and hence non-intentional. It is supreme bliss and non-dual. It is
to be realized.

It is only recently that modern science, particularly cognitive neurobi-
ology, has begun to explore the first three states of consciousness through
their neural correlates. However, there is no counterpart of the fourth and
unifying state in Western philosophy which stipulates 'intentionality' as a
necessary condition of consciouness in the sense that one cannot be conscious of nothing. There is here, therefore, a fundamental difference of approach between Western and Indian thought.

The neurobiological approach is a reductionist approach, and there are
strong arguments to show that reductionism holds only to a limited extent
even in the exact sciences.

6. The Limits of Reductionism

Machine designs are based on clever human concepts that exploit the
boundary conditions left open and arbitrary in fundamental natural laws.
For example, Newtons's laws do not specify when to throw a ball, in which
direction, and with what speed. This allows organisms to act as they wish
to a limited extent, and is necessary for life tasks including survival. Designs
of artefacts and machines are thus based on freely invented concepts that
exploit these inherent latitudes in the natural laws. They cannot therefore

be reduced to the laws. They have an autonomy of their own, though they need *some* physical basis for manifestation. Take, for example, the design of a clock to keep time. There are so many variations of it which exploit different physical systems. These physical bases can vary and be corrupted, but not the concepts or principles themselves. That is why a machine can be repaired. There is a certain parallel with the *Sānkhya* concepts of *prakruti* and *purusha* — the former is changeable, not the latter, but none of them can be reduced to the other. This must be even more true of the very ground of concepts, namely consciousness, and physical laws. Whether this duality can be synthesized is still an open question, though attempts at their unification have been made in the past in philosophical theories like *achintyavedāveda*.

We do so many things without knowing how we do them. For example, we can recognise a face in a crowd of thousands of people, but we cannot tell how. We can ride a bicycle, we can swim, but we cannot tell how. No amount of scientific knowledge can help us perform these acts. In fact, no such knowledge is even necessary. According to Michael Polanyi, this is because we have *tacit knowledge*. Such knowledge is more primary than any formal knowledge, for all formal propositional knowledge must be validated by tacit knowledge. It consists of tradition, inherited practices and their implied values, and prejudgments, and implies the scientist's personal participation in his knowledge. It is an indispensable part of science itself. Forensic scientists have developed computer programs to identify faces from a few clues. However, such programs would be useless unless they are first validated by human beings using their tacit component of knowledge. As Polanyi puts it, "We can know more than we can tell." We do not know how we do that. Perhaps we have acquired it through the process of performing life tasks. Even in the exact sciences like physics, "knowing is an art" which is logically necessary. This becomes more evident in the biological and social sciences. The paradigm of trustworthy, impersonal knowledge has split fact from value, science from humanity. There is a growing need now to replace this dehumanized, impersonal ideal of scientific detachment by an alternative ideal which draws on the personal involvement of the knower in all acts of understanding and restores a measure of wisdom. It is worthwhile to remember what T. S. Eliot wrote in his "The Rock" (1934):

> Where is the Life we have lost in living?
> Where is the wisdom we have lost in knowledge?
> Where is the knowledge we have lost in information?
> The cycles of Heaven in twenty centuries
> Bring us farther from GOD and nearer to the Dust.

I would like to end with a part of the famous *nāsadiya* hymn in the *Rig Veda* which captures the impenetrable mystery of all existence, and the very interrogative nature of reality itself (according to my philosopher friend Arindam Chakrabarti):

> But, after all, who knows, and who can say
> whence it all came, and how creation happened?
> The gods themselves are later than creation,
> so who knows truly whence it has arisen?
> Whence all creation had its origin,
> he, whether he fashioned it or whether he did not,
> he, who surveys it all from highest heaven,
> he knows - or maybe even he does not know.

References

1. Brentano, F. C., Psychology from an Empirical Standpoint, London: Routledge and Kegan Paul, (1911, 1973).
2. Born, M. & Wolfe, E., Principles of Optics. Cambridge: Cambridge University Press (1999).
3. Borges, C. V. S., Hor-Meyll, M., Huguenin, J. A. O. & Khoury, A.Z., Bell-like inequality for the spin-orbit separability of a laser beam, *Phys. Rev. A* **82**, 033833 (2010).
4. Davies, P. C. W. (2004). Quantum fluctuations and life, arXiv:quant-ph/0403017v1
5. Eckhorn, R., Bauer, R., Jordon, W., Brosch, M., Kruse, W., Munk, M., Reitboeck, H. J., Coherent oscillations: a mechanism of feature linking in the visual cortex? *Biol. Cybern.* **60**, 121–130 (1988).
6. Engel A. K. & Singer, W., Temporal binding and the neural correlates of sensory awareness. *Trends. Cogn. Sci.* **5**, 16–25 (2001).
7. Ekman, P., & Friesen, W., Constants across cultures in the face and emotion. *Jour. of Personality and Social Psych.* **17**(2), 124–129 (1971).
8. Ekman P., An argument for basic emotions. *Cognition and Emotion* **6**,169–200 (1992).
9. Ekman P., Facial expression and emotion. *American Psychologist* **48**, 384–392 (1993).
10. Ekman P., Levenson R. W., Friesen, W. V., Autonomic nervous system activity distinguishes among emotions, *Science* **221**, 1208–1210 (1983).
11. Gabriel, C., Aiello, A., Zhong, W., Euser, T. G., Joly, N. Y., Banzer, P., Fortsch, M., Elser, D., Andersen, U. L., Marquardt, C., Russell, P. S. J.& Leuchs, G., Entangling Different Degrees of Freedom by Quadrature Squeezing Cylindrically Polarized Modes, *Phys. Rev. Lett.* **106**, 060502 (2011).
12. Gerstner, W., Ritz, R. & Leo van Hemmen, J., A biologically motivated and analytically soluble model of collective oscillations in the cortex I. Theory of weak locking, *Biol. Cybern.* **64**, 363–374 (1993).

13. Ghose, P. & Samal, M. K. (2001, 2012). EPR Type Nonlocality in Classical Electrodynamics!, arXiv:quant-ph/0111119v1 22 Nov.

14. Ghose, P & Mukherjee A., Novel States of Classical Light and Noncontextuality, arXiv:1209.4026v3 [quant-ph] 20 Sep 2012, (submitted for publication).

15. Ghose, P. & Mukherjee A., Entanglement in Classical Optics, to be published in *Rev. Theor. Sci.*, American Scientific Publishers, USA (2013).

16. Holleczek, A., Aiello, A., Gabriel, C., Marquardt, C., Leuchs, G., Classical and quantum properties of cylindrically polarized states of light, arXiv:1012.4578v1 [quant-ph] (2010).

17. Husserl, E., *The Crisis of the European Sciences and Transcendental Phenomenology*, Evanston: Northwestern University Press (1970).

18. McCraty, R., Atkinson, M. & Tomasino, D., The Coherent Heart-Brain Interactions, Psychophysiological Coherence, and the Emergence of System-Wide Order. *Integral Rev.* **5** (2), 10-115, (2009).

19. Mohanty, J. N., *Transcendental Phenomenology: An Analytic Account*, Oxford and Cambridge, Massachusetts: Basil Blackwell (1989).

20. Polanyi, M., *Personal Knowledge: Towards a Post-Critical Philosophy*, University of Chicago Press (1958).

21. Polanyi, M., *The Tacit Dimension*, University of Chicago Press (1966, 2009).

22. Schneider, S. and Velmans, M., *The Blackwell Companion to Consciousness.* Wiley (1988).

23. Spreeuw, R. J. C., A Classical Analogy of Entanglement, *Found. of Phys.* **28** 361 (1998); *ibid.*, Classical wave-optics analogy of quantum information processing, *Phys. Rev. A* **63**, 062302 (2001).

24. Simon, B. N. *et. al.*, Nonquantum Entanglement Resolves a Basic Issue in Polarization Optics, *Phys. Rev. Lett.* **104**, 023901 (2010).

25. Varela F., Lachaux, J. P., Rodriguez, E. & Martinerie, J., The brainweb: phase synchronization and large-scale integration. *Nat. Rev. Neurosci.* **2** (4), 229–239 (2001).

26. Zurek, W. H., Pointer Basis of Quantum Apparatus: Into what Mixture does the Wave Packet Collapse?, *Phys. Rev. D* **24**, 1516–1525 (1981).

27. Zurek, W. H., and Paz, J.-P., Decoherence, Chaos, and the Second Law. *Phys. Rev. Lett.* **72**, 2508–2511 (1994).

Chapter 15

Consciousness — A Verifiable Prediction

N. Panchapakesan

University of Delhi, India

nargispanchu@gmail.com

Consciousness may or may not be completely within the realm of science. We have argued elsewhere[1] that there is a high probability that it is not within the purview of science, just like humanities and arts are outside science. Even social sciences do not come under science when human interactions are involved. Here, we suggest a possible experiment to decide whether it is part of science. We suggest that a scientific signal may be available to investigate the prediction in the form of an electromagnetic brainwave background radiation.

1. Levels of Explanation

Explanations can be broadly of three different types.[2]

(1) Bottom up,
(2) Same level or
(3) Top down.

Let us consider an example with the question: "Why is an aircraft flying?"

The scientific explanation is a bottom up one: Air molecules move at different speeds over the top and bottom wing surfaces to create a pressure difference that lifts the plane against gravity. This is the well known Bernoulli's principle. The explanation at the same level is: A trained pilot is flying it according to a scheduled time-table. The top down explanation is: The plane was designed to fly using technology to fulfill society's needs for transportation.

At each level, there are different types of explanations and different types of causality. Each explanation is partial and its preference is a choice based on context. Physics answers what and how; religion, philosophy answer why. The sum total of all my knowledge can be called a (or my) world view. Let us first understand a little bit about the nature of our world view.

2. World View

My thoughts and actions, my behaviour with other persons, my interaction with the world external to myself are decided in a very complex way by my conception of the world. This conception or view of the world, the basis for all my thinking and action, is often called World View of myself (or of the individual).

2.1. *Nature and Nurture*

My inheritance including genes, usually called nature, along with the influences of our surroundings, usually called nurture, have both contributed to my world view. Everyone of us has a world view, a collection of memory, knowledge, attitudes, values, vision and so forth. This is what guides our actions and our behaviour towards others. How is this world view formed?

2.2. *Values and Morals*

We can start with I, the individual who is at the receiving end. I, the thinker or the consciousness is at the centre of the inner world. I receive continually, **from birth**, sensory perceptions which after processing by the brain (by me) serve to define my world view. Free will or, capacity to make a choice is assumed. We assume that we can choose what we will receive, how we process and assimilate it, and how we act. (In any case, if there is no free will, then assuming it or denying it has no meaning. We can not assume or deny anything without free will.) In deciding or making a choice, one needs a value system, part of which is inborn and part comes from outside due to parental and other societal influences, including, of course, faith or religion (brain washing?).

2.3. *Outer world*

The ordering of one's perceptions defines a flow of time, a personal time which can be related to physical time of the outside world. Communication with the outside world by observing, speaking, listening, reading (use of language) coupled with logic enables the construction of the outer world.

I can, for example, infer that other persons also have an inner world and are doing things similar to me. We are also able to communicate our experiences to each other. However many experiences (including spiritual experience) of the individual can not be communicated adequately, which is of concern often. As is clear the centre of description of the inner world is oneself. Communicating with others enables me to realize the existence of an impersonal description of the reality outside me which may be called the outer world.

The *outer world* is the objective or impersonal world common to all human beings (us) which existed before my birth, holds me in it now and will continue to exist after my death. Though what happens after my death has no reality for me, I can visualise now, a world that may exist even in my absence. It is the world of physics and other sciences.

Emotion or feeling does not enter into the impersonal description of the outside world. Logic and scientific method (repeatability, falsifiability) are necessary. This world has a universal time and history. It is accessible to every individual through his or her perceptions.

2.4. *Natural and Social Sciences*

The distinction between natural and social science is very important. As the philosopher J. R. Searle[3] puts it "The distinction, rough as it is, between the so called 'natural' sciences and the 'social' sciences is based on a more fundamental distinction in ontology (essence of things), between those features of the world that exist independently of human attitudes, **like force, mass, gravitational attraction and photosynthesis**, on the one hand, and on the other, those whose existence depends on human attitudes like **money, property, marriage and government**. There is a distinction, to put it in very simple terms, between those features of the world that are observer-independent and those that are observer-dependent. Natural sciences like physics, chemistry and biology are about features of nature that exist regardless of what we think, and social sciences like economics, political science and sociology are about features of the world that are what they are because we think that is what they are."

The observer dependence that Searle talks about is different from the observer dependence in natural sciences especially in quantum mechanics. Searle is referring to the dependence of concepts and ideas on culture, society and state.

3. Error Magnitudes Differentiate Social and Natural Sciences

One can have further gradations in the social sciences depending on the nature of subjective involvement, as measured by the uncertainties in observations. The large uncertainties make individual observations meaningless. We have to have large number of observations and use statistical methods for drawing conclusions. Smaller the sample larger is the uncertainty in the observation or the prediction. The prediction of what one person will do is fraught with very large uncertainties, while statistical predictions like, "25% will act in a particular way", are somewhat easier to make. Actual observation may show it to be between 20 to 30%.

Spirituality is deep inside the inner world. There is no conflict between it and science which is in the outer world, as long as they stick to their territories. Consciousness can thus be looked at from different levels; bottom up being scientific, top down being intuitive.

Essence of Spirituality is the idea of unity of the whole universe. S. Radhakrishnan, the Indian philosopher, statesman says: "We have also the conviction of the unity of the universe. For the intellect the unity is only a postulate, an act of faith. For the spirit, the harmony is the experienced reality."[4]

Albert Einstein has also written about spirituality. He writes: "All religions are a varying blend of both types (religion of fear and moral religion) $\cdots$ $\cdots$ on the higher levels of social life the religion of morality predominates $\cdots$. But there is a third stage of religious experience which belongs to all of them (religions), even though it is rarely found in a pure form: I shall call it cosmic religious feeling. It is very difficult to elucidate this feeling to anyone who is entirely without it, especially as there is no anthropomorphic conception of God corresponding to it. Individual existence impresses him as a sort of prison and he wants to experience the universe as a single significant whole. $\cdots$ Buddhism $\cdots$ contains a much stronger element of this." (Ideas and Opinions)[5]

Golden Rules including Advaita: The basis for guidance of our actions can be put in the form of simple rules like:

- Do unto others as you would that they do unto you. (Feynman's favorite, he thought it was beyond science.)
- Individual self and the Universal self are the same.
- God (not anthropomorphic) is within everyone. Everyone is unique and equal at the same time.

Using the rule gives guidance for moral decisions.

3.1. *Possible Signals*

We start with the hypothesis of unity of all selves (consciousnesses) and try to see if there are any experimental signals for it. We can ask if there is a medium or a physical field connecting all selves. If the medium is the electromagnetic field the we could try to observe if we are soaked in a sea of electromagnetic radiation, like the cosmic microwave background radiation (CMBR) in space.

3.2. *Possible Frequencies*

Brain is known to emit e.m. waves (of wave lengths $\sim$ Earth's radius): EEG

- Delta $1-4$ Hz, $\cdots$ Deep sleep,
- Theta $4-7$ Hz   Drowsiness,
- Alpha $7-13$ Hz   Relaxed but alert,
- Beta $14- 30$ Hz   Highly alert, focussed,

One could start looking in the range $1-30$ Hz, and setting upper limits of intensities like for gravitational radiation. Experiments may be able to verify this idea.

3.3. *Alternatives*

If no BBR (Brainwave background radiation) is found, are there other ways of reconciling science and spirituality? It may be possible that quantum theory may provide a demarcation between inner and outer world. The inner self may be in the entangled state and outer self in the collapsed state of the wave function. Gödel-Turing theorems already separate the algorithmic from non-algorithmic systems. This kind of resolution is a far deeper one and depends on developments in quantum measurement theory and the experiments that go along with it.

4. Summary

(1) Can there be knowledge or awareness outside science?
(2) There are approaches other than bottom-up or reduction of everything to its constituents. Top-down, for example.

(3) Our world view spreads from spirituality to natural science and indicates areas which are not part of natural science.

(4) These areas require intuition and lead to plurality and lack of unanimity.

(5) There is however a large consensus on a feeling of unity and relatedness among all things, especially human beings.

(6) If humans are connected up, it could be in physical or some other space.

(7) If the unity is in physical space, then we suggest that the intervening medium or field can be electromagnetic.

(8) Brain emits electromagnetic radiation at a very low frequency. If everything is united, there could be cosmic background radiation connecting all brains. This suggestion could be experimentally tested. We suggest doing such experiments.

References

1. N. Panchapakesan, "Need for Change of Paradigm in Science Teaching", *Econ. Pol. Weekly* Vol.-XLIII, No.17, April 26, (2008).
2. G.F.R. Ellis, "Physics and the Real World", *Physics Today*, July, p 49. (2005).
3. J. R. Searle, "End of the Revolution", *The New York Review*, February 28, pp 33-36 (2002).
4. S. Radhakrishnan, *An Idealist View of Life*: being the *Hibbert Lectures* for 1929. George Allen & Unwin, London (1932).
5. Albert Einstein, *Ideas and Opinions*, Crown Publishers, New York (1954).

Chapter 16

Gödel, Tarski, Turing and the Conundrum of Free Will

Chetan S. Mandayam Nayakar and R. Srikanth

Poornaprajna Institute of Scientific Research, Sadashivnagar, Bangalore
and
Raman Research Institute, Sadashivnagar, Bangalore
srik@poornaprajna.org

The problem of defining and locating free will (FW) in physics is studied. On the basis of logical paradoxes, we argue that FW has a meta-theoretic character, like the concept of truth in Tarski's undefinability theorem. Free will exists relative to a base theory if there is freedom to deviate from the deterministic or indeterministic dynamics in the theory, with the deviations caused by parameters (representing will) in the meta-theory. By contrast, determinism and indeterminism do not require meta-theoretic considerations in their formalization, making FW a fundamentally new causal primitive. FW exists relative to the meta-theory if there is freedom for deviation, due to higher-order causes. Absolute free will, which corresponds to our intuitive introspective notion of free will, exists if this meta-theoretic hierarchy is infinite. We argue that this hierarchy corresponds to higher levels of uncomputability. In other words, at any finitely high order in the hierarchy, there are uncomputable deviations from the law at that order. Applied to the human condition, the hierarchy corresponds to deeper levels of the subconscious or unconscious mind. Possible ramifications of our model for physics, neuroscience and artificial intelligence (AI) are briefly considered.

1. Introduction

Informally speaking, FW is the power to choose one alternative from many. We think that we have it. Yet its existence and nature have been debated for over two thousand years by philosophers, scientists and theologians.[1,2] Free will is, arguably, a familiar stranger. One basic difficulty here is its paradoxical nature: FW incorporates two opposing notions: freedom and control. On the one hand, freedom suggests indeterminism; while control suggests intent and determinism. From this perspective, FW is an oxymoron. There are some easy questions to ask about it that are hard to answer:

(1) *What is FW?* Physicists typically treat it as unpredictability or uncorrelatedness with some past data. By this criterion, uncontrolled or unreasonable behavior would qualify as free, and thus does not work. On the other hand, if control and predictability are the requirements, then the altruism of a saint and the selfish actions of a materialist are equally predictable, though intuitively, the former is the one we would deem free. The relevant question here is: how to define FW?

(2) *Does it exist is the world?* Introspection suggests that we are free in the sense that we could have done other than what we did. But unless FW is well defined, this intuitive feeling may well be illusory!

(3) *Is it compatible with the laws of physics?* The answer to this would depend on those to the above two questions!

(4) *What is its neurological basis, if FW exists?* Here again, the answer depends on the answer to the above three.

This article addresses the first three questions above. The last is dealt with by us elsewhere.[3] There are three broad metaphysical positions on FW: *Compatibilism*, which holds that determinism is compatible with FW. A person may choose freely and yet an omniscient being may possess foreknowledge of that choice. The opposite view is *incompatibilism*, according to which FW and determinism are incompatible. Two divergent incompatibilist views are *hard determinism*, which rejects FW in favor of determinism, and *libertarianism*, which rejects determinism in favor of FW. To the determinist, free will is at best illusory. The variety of FW that arguably poses the greatest challenge is the liberatarian one, and it is this that we consider in this work.

The remaining article is structured as follows. In Section 2, we briefly review the current status of free will in quantum mechanics. In Section 3, we present the Weak Free Will paradox, which is concerned with question of the compatibility of FW with the laws of physics. A recapitulation of a model of FW we developed in Ref.[4] in response to the Weak Free Will paradox is presented in Section 4. The Strong Free Will paradox, the argument that the only causal primitives in Nature are indeterminism and indeterminism, is presented in Section 5. In response to the latter paradox, FW is proposed in Section 6 as an infinite causal hierarchy of meta-laws, where higher-order laws can cause deviations in lower-order ones. The orders of the hierarchy are then differentiated by different levels of uncomputability in Section 7. A justification for this association is presented in Section 8. The fuller implications for neuroscience, physics and AI are discussed in Section 9. Finally, we conclude in Section 10.

2. Quantum Mechanics and FW

Quantum mechanics, by introducing the new paradigm of intrinsic randomness provided fresh impetus to the FW debate. At first it seems that quantum indeterminism provides room to accomodate free will. Yet, randomness also means loss of control.[5] Two kinds of freedom are sometimes discerned: that of experimenters who choose measurement settings, and of particles that decide the measurement outcome.[6] T' Hooft[7] proposes identifying FW with freedom to modify the initial state of a system.

Reconsideration of randomness in light of quantum nonlocality has provided new insights. Gisin[8] has argued that the Many-Worlds interpretation[9] of quantum mechanics is incompatible with quantum nonlocality and the existence of FW. The Conway-Kochen FW Theorem[10,11] proves that given the violation of Bell-type inequality and relativistic time-ordering, the assumption of freedom of choice of settings by observers implies a similar 'free will' on the part of particles, i.e., independence from all past information (with 'past' defined as the past light cone or all complement of the future light cone). Particle outcomes are then acts of creation or becoming rather than pre-determinate values.[12] The claim[10,11] that the freedom of experimenters and the violation of Bell-type inequalities rule out covariant stochastic dynamical explanations of quantum nonlocality was disputed by others.[13,14] A reconciliation of these two positions can be brought about by the observation[12,15] that whereas 'collapse' of the nonlocal wavefunction leading to correlated outcomes is not covariant, still the 'cloud of future events' is. Hall[16] has refined the notion of FW discussed in the Free Will theorem, by obtaining a trade-off between the freedom of observers or measurement independence and the freedom of particles or indeterminism in measurement outcomes.

Suarez[17] has argued that assuming freedom of the experimenter in a measurement configuration where two observers perform a Bell inequality test,[18] such that each considers her/his own measurement as having been performed first, then time-ordering of physical causation must be given up. Outcomes are then controlled by an immaterial agency outside space-time that determines the order of outcomes without affecting the probabilities, given by the Born rule. By this argument, FW and randomness can coexist.[19]

3. The Weak FW Paradox

Consider world W governed by *deterministic* physical law, L. All objects in W evolve according to L. The pair (W, L) constitute theory $\mathcal{T}$. By determinism, the physical state ψ_t at time t of an inanimate agent (e.g., a particle) and the physical law determine the state ψ_{t+1} at subsequent time $t + 1$: $\psi_{t+1} = L(\psi_t)$. Free will seems to entail for a conscious agent X in W, that in general,

$$\psi_{t+1} \neq L(\psi_t), \tag{1}$$

meaning that X's behavior is not captured by L. We may consider replacing L by a probabilistic law L' such that

$$\Psi_{t+1} = L'(\Psi_t), \tag{2}$$

where Ψ_{t+1} is a random variable representing probabilitistic evolution, characterized by mean μ and standard deviation σ. By the Law of Large Numbers, the sample mean $\langle \Psi_{t+1} \rangle^{(n)}$ of the random variable Ψ_{t+1}, over n trials satisfies

$$\lim_{n \to \infty} \Pr\left(\left| \langle \Psi_{t+1} \rangle^{(n)} - \mu \right| > \epsilon \right) = 0. \tag{3}$$

In other words, it is exponentially unlikely that the agent X will, over many trials, choose *atypical* sequences. There is thus long-run determinism and on that scale, lack of freedom.

Thus libertarian FW is compatible with neither a deterministic nor indeterministic physical law. This is the Weak Free Will paradox. We conclude that, libertarian FW, if it exists for an agent X, must be a supra-physical commodity that causes *deviations* from the laws of the world physical W in which X (physically) lives. Causality is no longer closed under physics in a world where FW exists. Otherwise– if free-willed action were explainable by L–, then there is no deviation by definition, and FW must be illusory.

The *quantum homeostatis hypothesis*[17] is another response to the Weak Free Will paradox. According to it, a conscious agent may exert FW by controlling the order of outcomes, without contradicting the probability law Eq. (2). In order not to deviate from the long term distribution required by L', periods of deliberate, conscious action must be compensated by other periods of uncontrolled behavior (e.g., sleep).

4. Free Will as the Power of Mind over Matter

A proposal that incorporates will-driven deviations from the physical law, as required in our resolution of the Weak Free Will paradox, is the Freedom-Understanding-Nature (FUN) model.[4] An agent is assumed to be equipped with two possibly opposing faculties: Nature (N), determined by physical law L via brain (limbic) dynamics, genetic proclivities and instinctual drives, and Understanding (U), determined by higher cortical faculties of morality and compassion. Here N is a manifestation of L from which will-driven deviations may occur. For example, N may manifest as a psychosomatic desire that urges the agent to procure a physical object, while U may advise restraint. Freedom (F) is an immaterial resource that empowers X to deviate choice C away from the option urged on by N towards the one advised by U. This deviation cannot be captured by a Hamiltonian dynamics or a mechanical model, since these exist within L. The *base-law* of the deviation corresponds to the known physical law (here N), but the parameters of deviation (here φ and possibly U) are associated with the *meta-law*. Together the meta-law (which encompasses L) and parameters associated with it consistute the *meta-theory*.

Free will is the power to conform to U, deviating from N if necessary. A simple mathamatical model: suppose agent X is faced with a two-valued choice "0" or "1". U and N are represented by random variables $\hat{U}$ and $\hat{N}$ that take values in dichotomic choice space of $\Omega \equiv \{0, 1\}$. Let U advise option "0" whereas N urges X towards option "1" with a 'strength' ν. The eventual choice C of the agent is represented by random variable $\hat{C}$, which is a convex combination of $\hat{U}$ and $\hat{N}$, with the weight assigned to U determined by the freedom parameter φ (where $0 \leq \varphi \leq 1$), representing F. Thus:

$$\hat{C} \equiv (1 - \varphi) \begin{pmatrix} 1 - \nu \\ \nu \end{pmatrix} + \varphi \begin{pmatrix} 1 \\ 0 \end{pmatrix} = \begin{pmatrix} \varphi + (1 - \varphi)(1 - \nu) \\ (1 - \varphi)\nu \end{pmatrix}. \qquad (4)$$

In Eq. (4), the larger is φ, the more can C deviate from N towards U. If freedom is maximal, i.e., $\varphi = 1$, then the agent behaves deterministically (mathematically, $\hat{C}$ is pure), demonstrating that FW is distinct from unpredictability in this model. We note that $\hat{C}$ becomes more random as φ drops from 1.

Free will Φ is quantified as a measure of closeness of $\hat{U}$ to $\hat{C}$ normalized by the distance of $\hat{U}$ from $\hat{N}$.

$$\Phi = \left(\vec{C} \cdot \vec{U} \right) |\vec{U} - \vec{N}|, \qquad (5)$$

where $\vec{C}$ $(\vec{N})$ is a vector representation of $\hat{C}$ $(\hat{N})$ and $|\cdot|$ denotes trace distance. By definition, $0 \leq F \leq 1$. For Eq. (4), we find

$$\Phi = \frac{\varphi}{2}(\varphi + (1 - \varphi)(1 - \nu))|1 - 2\varphi|. \tag{6}$$

We may thus think of FW Φ as innate freedom F (represented by φ) *expressed* physically when there is a conflict between U and N. The greater the conflict, the greater the potential for this expression. Here innateness means that F is a part of the meta-theory (or higher-order theory), while its expression is at the base, physical level.

5. The Strong FW Paradox

That libertarian FW produces deviations from the base law, only tells how FW acts. It does not clarify what this commodity is. We now attempt to define FW as the resource that can produce such deviations. The Strong Free Will paradox, given here, is concerned with the question of whether these deviations themselves are lawful according to a higher law. Suppose agent X is faced with a situation that presents a choice. By our resolution of the Weak Free Will paradox, in free choice, physical causation fails in W in that X may deviate from the physical law L. X has FW relative to W. The parameters of this deviation are not governed by L, and therefore are objects that live in the extended world W^+. One now considers whether these parameters themselves are lawful or free-willed. For agent X to have genuine FW, these deviation parameters should have been freely chosen.

One possibility is that there there are deterministic laws L^+ in W^+ that explain the causes that produce the deviation on the base law L. We then have:

$$\psi_{t+1}^+ = L^+ \left(\psi_t^+ \right), \tag{7}$$

where ψ_j^+ is the extended version of ψ_j, obtained by adding elements from W^+. For example, suppose W^+ includes the physical objects in W as well as other more abstract, mental objects. Then ψ_t represents the state of agent X's body at time t, then ψ_t^+ is that of his body-mind at that time. Here ψ_{t+1} is implicitly determined by L^+ acting on the enhanced state ψ_t^+. In this case, there is no FW in W^+, and in an absolute sense, no FW at all, since the deviations from L are deterministically determined by the higher law L^+. Another possibility is that there is a probabilistic law $L^{+\prime}$ that governs the W^+ objects, but here again there is no FW, as argued in Section 3. By the Weak Free Will paradox, which is adapted to W^+ agency, there is FW only if there can be deviations from law L^+.

The causes (or parameters) of these deviations are objects that live in a yet higher world W^{++}. It is clear that the above argument *relativizes*, i.e., one can recursively apply the above argument, shifting the level from W^+ to W^{++}, and beyond. If this recursion is finite, then although there is FW in the sense of the Weak Free Will paradox till that causal depth, still there is none beyond. What remains just above that depth is either determinism or indeterminism. In other words, the Strong Free Will paradox is the argument that the only fundamental causal primitives are determinism and indeterminism. There is no room for any primitive like FW, which is seen ultimately to be a hierarchical interplay of these two primitives.

The quantum homeostatis hypothesis[17] and the FUN model,[4] which provide a resolution to the Weak Free Will paradox, do not appear to get past the Strong Free Will paradox. In the former case, the Strong Free Will argument applies to the resource by which an agent controls the ordering of outcomes, and in the FUN model, it applies to the freedom paramter φ.

6. Free Will as Infinite Meta-Theoretic Construct

We propose that the response to the Strong Free Will paradox is to simply embrace it! We suggest that it is in the nature of Consciousness that it can support the above causal recursion to any depth. For convenience, let the base (physical) laws L now be denoted by $L^{(0)}$ and the corresponding physical world W in which they operate, by $W^{(0)}$. The usual laws of physics are order-0 laws, which govern the behavior of order-0 objects, which are physical objects, denoted $\lambda^{(0)}$, such as the physical component (body) of sentient agents like humans. The physical state of such a being is denoted $\psi^{(0)}$.

Volition- or will-driven deviations (if they exist) from the order-0 laws are caused, according to Eq. (1), by order-1 objects in L^+, which we denote now by $L^{(1)}$. These objects, denoted by $\lambda^{(1)}$, live in the larger world W^+, denoted by $W^{(1)}$. We have $W^{(0)} \subset W^{(1)}$, $\lambda^{(0)} \in W^{(0)}$ and $\lambda^{(1)} \in W^{(1)} - W^{(0)}$. The extended state of the free-willed being X is $\psi^{(1)} \in W^{(1)}$ such that $\psi^{(0)} = \psi^{(1)} \cap W^{(0)}$. Similarly, will-driven deviations from $L^{(1)}$ dynamics are attributed to causes $\lambda^{++} \equiv \lambda^{(2)} \in W^{++} \equiv W^{(2)}$. The law $L^{(1)}$ is a meta-law for $L^{(0)}$ in that it is law that is about and governs $L^{(0)}$. The objects in $W^{(1)}$ and the axioms of $L^{(1)}$ thus constitute a meta-theory $T^{(1)}$ for the base theory $T^{(0)} \equiv (W^{(0)}, L^{(0)})$.

At each level j (a positive integer), the pair $(W^{(j)}, L^{(j)})$ constitute the order-j meta-theory $\mathcal{T}^{(j)}$. Formally, this system of meta-theories $\mathcal{T}^{(j)}$, can be continued indefinitely. The meta-theoretic construction of FW is the assertion that at each level j, FW at that level is expressed as the possibility of deviation from $L^{(j)}$ caused by order-$(j+1)$ objects $\lambda^{(j+1)} \in W^{(j+1)}$. *Relative* FW exists at level K if there are deviations from all $L^{(j)}$ with $j < K$. If there are no deviations in $L^{(K)}$, then there is no FW beyond order-K. We propose the existence of *absolute* FW exists in the sense that relative FW exists for all finite K. In other words, no matter how deep we go tracking down causes, the causes of these causes, the causes of causes of causes, and so forth, although our explanatory power may increase, still there remain unexplained deviations.

We note that determinism and indeterminism lack this meta-theoretic character in that they can be described within a given level (say L) of the dynamics. Free will then is a new causal primitive, in addition to determinism and indeterminism. But although it is not logically necessary for indeterminism to be meta-theoretic, still it may be a function of higher-order causes. It differs from free will at the same order in not being correlated with the immediately higher order meta-theory. Thus indeterminism is just freedom without the will. Quantum fluctuations are then order-0 freedom, while sentient beings who deviate from $\hat{N}$ in the FUN model possess (at least) order-0 free will.

If no such correlated parameter is evident in the meta-theory, then there is freedom alone, and to ensure that causality holds, we conclude that the causes lie in meta-theoretic depths farther down. This is a subtle but socially significant difference between the free will of sentient agents and the freedom of particles. Although there is no logical necessity to do so, we propose that all indeterminism is freedom in the above sense.

A problem that merits further study is the question of how the different meta-theories $\mathcal{T}^{(j)}$ may be demarcated. One may consider that $\mathcal{T}^{(0)}$ corresponds to predicate logic, and $\mathcal{T}^{(j)}$ ($j > 0$) correspond to higher-order logics.[20] An example of a demarcation that doesn't work is the proposal that $\mathcal{T}^{(0)}$ is classical mechanics and $\mathcal{T}^{(1)}$ quantum mechanics, the idea being that quantum mechanics causes deviations from classical behavior. We venture that quantum mechanics, and thus also classical mechanics, belong in $\mathcal{T}^{(0)}$, the base physical theory. Intuitively, this is because quantum mechanics can be considered as a modification over an indeterministic generalization of classical mechanics. Although this modification is responsible for peculiarities like quantum nonlocality, it does not seem to endow quantum mechanics with more explanatory power, as we clarify below.

7. Computability and Causality

We suggest that the different levels of theory $T^{(0)}$, the meta-theory $T^{(1)}$, the meta-meta-theory $T^{(2)}$, etc. are distinguished in different levels of *computability* that they support. Computability, which is the capacity to effectively solve a problem, is a topic studied in mathematical logic and computer science. Computability theory is concerned with the question of whether an algorithm or effective procedure exists to solve a problem. It is known that there exist algorithmically uncomputable problems, e.g., the halting problem for computer programs (formally: Turing machines).[21] One can conceive qualitatively more powerful computers (formally: oracle machines) that can solve these problems, but must contend with their own halting problem. One can continue this exercise, to build a hierarchy of ever more powerful oracle-machines and ever harder uncomputable problems. We propose that the different $T^{(j)}$'s correspond to the different rungs in this hierarchy.

We can now make precise the sense in which FW is said to exist in this model. In computation theory, an *oracle* for problem A is a hypothetical device that given any instance of A returns the answer in finite time. If a Turing machine with access to this oracle can solve every instance of problem B, then B is Turing-reducible to A. In other words, if A is computable, then so is B. If similarly, A is reducible to B, then A and B are equivalent. A Turing-degree is the set of all problems that are mutually Turing-equivalent.

The set of all computable problems is a Turing degree, which quantifies the level of its algorithmic unsolvability. A higher Turing degree is the set of problems Turing-equivalent to the halting problem, for example, the problem of whether a statement can be proved from the axioms of set theory. This latter set is not solvable (with Turing machines). One can construct higher Turing degrees by relativizing the proof of the halting theorem for Turing machines to Turing machines with oracles.

Thus the different meta-theories correspond to different *Turing degrees*, with a higher-order theory being of higher Turing degree. The move from $T^{(j)}$ to $T^{(j+1)}$ is equivalent to a *Turing jump*,[22] i.e., produces a unit shift in the Turing degree.

By this criterion, since the quantum version of Turing machine is not more powerful than a classical machine from a computability perspective (and even computational complexity perspective), quantum mechanics belongs to the same-order theory as classical mechanics, namely $T^{(0)}$. On the other hand, a deterministic hidden variable theory for quantum mechanics is a likely candidate for $T^{(1)}$. For one, it can deterministically explain deviations from classical mechanics in terms of hidden variables. Further, the ability to manipulate this subquantum information leads to super-Turing power in a computational complexity sense (solving NP-complete problems in polynomial time).[23]

Our assertion of absolute FW above can now be re-stated thus: it exists in the sense that the problem of determining an agent's free choice is of infinite Turing degree, i.e., for every oracle machine at finitely high order, there will be uncomputable residual deviations in a model of free choice. In other words, FW is not just uncomputable, but infinitely so!

The first hint that the this linking between causality and computability is an appropriate direction to proceed is the observation that the notion of *proof* in Gödel's incompleteness theorem[24] and of *truth* in Tarski's undefinability theorem,[25] are, in some ways, similar meta-theoretic concepts. Gödel incompleteness may be considered as an avatar of Turing uncomputability,[21] as we show in Section 8.

8. Truth, Proof and Freedom

Imagine a human agent X equipped with the usual cognitive apparatuses of a physical brain and something else called mind. The physical aspect of X lives in world $W^{(0)}$ governed by physical law $L^{(0)}$, while the higher aspects of $\psi^{(j)}$ are attributed to the mind entity. Now $T^{(0)}$ naturally determines a model of computation, and correspondingly the limits of computational complexity and computability, in the world $W^{(0)}$. If $T^{(0)}$ is the physical world, then the relevant model is Turing machines (TMs) or their equivalent. All TMs, including the TMs representing agents X, Y, Z et al. that live in $W^{(0)}$, can be enumerated according to a fixed scheme, e.g., Gödel numbering or Turing numbering. We denote the corresponding numbers $\underline{X}, \underline{Y}, \underline{Z}$ and so on.

We now present a proof of Gödel's theorem via the uncomputability of a problem that is equivalent to the halting problem.[21] Suppose algorithm $\mathcal{A}$ exists that can predict in finite time agent X's binary choice upon input i:

$$\mathcal{A}(\underline{X}; \underline{i}) = \begin{cases} 0 \iff X(i) = 0 \\ 1 \iff X(i) = 1 \end{cases}, \tag{8}$$

i.e., the machine outputs 0 (1) if X outputs 0 (1) acting on input i. We construct the following program:

$$\mathcal{R}(p) = \begin{cases} 1 \iff \mathcal{A}(p; p) = 0 \\ 0 \iff \mathcal{A}(p; p) = 1, \end{cases} \tag{9}$$

where p is a positive integer. By construction, $\mathcal{R}$ exists if $\mathcal{A}$ does. Applying $\mathcal{A}$ to $\mathcal{R}$ we have from Eqs. (8) and (9):

$$\mathcal{A}(\underline{\mathcal{R}}; p) = \begin{cases} 1 \iff \mathcal{A}(p; p) = 0 \\ 0 \iff \mathcal{A}(p; p) = 1. \end{cases} \tag{10}$$

We obtain a contradiction when we set $p = \mathcal{R}$, which is an instance of the diagonal argument, first pioneered by Cantor. The conclusion is that a general, sound predictive algorithm $\mathcal{A}$ is logically impossible because it would help create an algorithm that is so powerful that, knowing its own future, it could act contradictory to its own prediction.

Therefore, if the system F of reasoning embodied by TMs in consistent, then to avoid the contradiction, $\mathcal{A}(\underline{\mathcal{R}}; \underline{\mathcal{R}})$ loops infinitely. Therefore, $\mathcal{R}(\underline{\mathcal{R}})$ is undecidable within F– a Gödel sentence for the system. How do we, as human beings reading this, know this truth of an algorithmically undecidable statement like $\mathcal{R}(\underline{\mathcal{R}})$? Can the human mind therefore be more powerful than TMs? According to a much debated anti-mechanist view due to Lucas[26,27] and Penrose,[28,29] it is.

An objection[30] is based on Gödel's second incompleteness theorem. The reasoning that led us to the claim that "If F is consistent, then $\mathcal{R}(\underline{\mathcal{R}})$ is non-terminating" can itself be programmed into the action of a TM. Thus if F is in fact consistent, then the proof of its own consistency must be uncomputable, in order that $\mathcal{R}(\underline{\mathcal{R}})$ be unprovable in F. In this objection to the anti-mechanist view of the mind, even if human agents were consistent, they would be unable to establish this, and thus attain to a meta-mathematical comprehension of what TMs can't. To this view, the following counter-objection may be given: Suppose a large number n of persons toss a fair coin each, assuming that humans are consistent if a head turns up, and that we are inconsistent otherwise. Then it follows that if it so happens that human beings are consistent, the understanding that $\mathcal{R}(\underline{\mathcal{R}})$ is non-terminating would make the mind will be more powerful than TMs in the case of those people in the group for whom ($n/2$ in number, with exponentially high probability) a head turns up! Clearly, we wouldn't want the power of the mind to depend the whim of a coin! (On the other hand, if humans happen to be inconsistent, then the above argument fails.) Thus it seems that objections based on consistency of F do not satisfactorily undermine the anti-mechanist view.

Another criticisim[31] is that it may not be possible to humanly construct a Gödel sentence for something so complicated as the TM representing a human being. However this objection seems to assume that the human may in principle be more powerful than a TM, but not in practice. The claim for the disputing the anti-mechanist view should, it would seem, not depend on limitations imposed on human beings like mortality, emotionalism, fatigue, etc., unless it can be demonstrated that such a factor will necessarily (logically) intervene to prevent a human from computing the Gödel sentence in question.

Yet another objection to the anti-mechanist view is due to Whiteley,[32] which is that humans would themselves be prone to a Gödel sentence, not unlike TMs. Consider the following predicate formula, formalized suitably:

$$P(\xi) \equiv \text{``}\xi \text{ cannot consistently believe this statement.''}$$

Now $P(X)$ is true or false. If true, then X does not believe $P(X)$ when presented with it, making him *incomplete* for disbelieving a true statement. If $P(X)$ is false, then $\neg P(X)$ is true; in other words, he *believes* a false statement, making him *inconsistent*. To avoid inconsistency, we select for him to be incomplete. This is just an informal version of Gödel's theorem, with disbelief[33] self-referenced, rather than unprovability (Actually, presenting $P(X)$ to X creates a double self-reference: in X and in the resulting statement[34]). According to this objection, humans have similar limitations as those attributed to machines on basis of Gödel's theorem.

$P(X)$ can be consistently believed by agent Y, and $P(Y)$ by X. Now imagine that Y believes $P(X)$ and informs this statement to X. As predicted X is unable to believe $P(X)$, being consistent. When Y asks him if he believes $P(X)$, X replies in the negative. When Y asks him why, X says that he cannot say exactly why, but somehow $P(X)$ feels unconvincing. Now Y happens to be a mathematician, and explains Gödel's theorem to X enlightening X as to why X is unable to believe $P(X)$. X is now able to reflect upon the nature of his understanding and the limitations imposed by the theorem. He trusts Y enough to know that Y wouldn't believe $P(X)$ unless it were true. He then has an 'Ah ha' moment, and decides that $P(X)$ is true after all. His brain still disbelieves $P(X)$, but somehow he knows that $P(X)$ is true, though maybe he can't help acting as if it were false. If X could *not* have this Ah-ha moment, he could well be a machine, and in social terms, he must be some kind of zombie or dull-witted person, not a normal human! Thus, the power of intuitive grasp seems to be that X can supply his own meta-theory on the fly, and it is this power to meta-jump that seems to come from nowhere but the nature of consciousness, and divides man and machine!

In other words, X is able to *meta-straddle*: he can both believe and disbelieve $P(X)$ simultaneously, without being inconsistent because these states occur at two different layers: the disbelief at $W^{(j)}$ (possibly $j = 0$) and the belief at $W^{(j+1)}$. (Here we may recollect moments where our eyes/brain are deluded by an optical illusion which we know is illusory; or when we know that something is false at a higher level, and yet can't help believing it!) This conclusion is consistent with our demarcation of the meta-theories $T^{(j)}$ in terms of uncomputability occurring in the theory.[34]

Now it is conceivable that the reasoning that led us to the conclusion that if X intuits the Gödel sentence, and then meta-straddles, can itself be formalized or equivalently mechanized somehow, for example having a multi-tape TM with a denial on the lower tape, and acceptance on a higher tape that accesses an oracle. So this order-1 oracle machine will be our new model for agent X. However, one can construct a Whitely sentence $P^{(1)}(X)$ for X modeled as an oracle-1 machine, and repeat the above argument, for X perceiving the truth of this sentence. Relativizing this procedure, we construct ever more powerful oracle machines, going up the jump hierarchy, and iterating farther into transfinite ordinals, and even uncountable regular cardinals.[22] In each case, X modeled as these machines still makes the intuitve jump, because if it didn't, we would call it a (higher-order) zombie. Whereas the very character of a machine must be changed as we climb up the jump hierarchy, the consciousness of X seems essentially to be the same thing, as it jumps up or hops outside endlessly. It has the magical infinitely, meta-expansive plasticity that seems to be the essence of conscious understanding, as against mechanized assimilation by an automaton.

9. Mind and Magic: Possible Ramifications for AI, Physics and Neuroscience

Without going into details, we may assume that the relation between the physical aspect and the mind of human agents is causal both ways, i.e., that each influences the other. We may deduce that there are oracles located in the brain, that serve as gateways through which $\lambda^{(1)}$ and higher influences can flow into the physical system, to produce deviations from $L^{(0)}$ laws, giving rise to free will in the physical plane. Since our conscious actions are arguably not more powerful than TMs, it follows that the influence of these brain oracles must occur at a subconscious or unconscious level. Therefore, the only failsafe way to access this super-Turing computational capacity involved in a person's free choice is to present the person with the choice, and allow him to react spontaneously. To use a Biblical allusion, there was no way to compute what Adam's reaction would be to the injunction forbidding him to eat the apple in the forest of Eden, except by putting him physically in the forest and forbidding him! From a computational perspective, every instance of free choice is a miracle! These observations have various ramifications discussed below.

Our model leads naturally to a Cartesian mind-body dichotomy, where the mind is indefinitely layered into increasing levels of uncomputability. Since the conscious body/mind is no more powerful than TMs, the entirety of the human consciousness cannot be consciously grasped through a finitary reasoning process, and must be done intuitively. From the perspective of the physical world, the actions of a free-willed agent are not always predictable, without being random. Mathematically, free choice can be described probabilistically as in Eq. (4) of the FUN model, where probability arises through *unknowability*. Any AI algorithm or neural network system, being no more powerful than Turing machines, can never in general simulate a free choice of human or presumably even an animal or, in general, any entity 'vivified by a soul'. Thus Consciousness, of which intuitive thought and FW are aspects, must also be also unsimulable, in contradistinction to the premise of Strong AI.

It seems possible that phenomena like savantism or the pattern recognition in the brain make use of the higher computational capacity attributed to the subconscious mind in our model. There must be something special in the structure of the brain that allows an interface with a subconscious oracle, a point that we discuss elsewhere.[3] We venture that the above phenomena are windows on that structure.

The free-willed human being is not an isolated entity, but part of the universe. If FW has the above implications for the physics of the human brain, this must have some implications, significant or otherwise, for the physics of the universe at large. If we extrapolate the above of hierarchy of laws to the large scale, then the grand unified theory (GUT), when discovered, will be simply be the tip of an indefinite hierarchy of meta-theories. We can then ask whether a notion of FW can, in a sense, be considered on a larger, cosmological scale. Perhaps spontanenous symmetry breaking and even the Big Bang, were possibly free choices by the 'mind of the universe'. Finally, given the greater computability supported the higher-order laws, it is impossible for physical laws to 'compute' them into existence. Thus the Big Bang may well be the 'ground zero' of a Bigger Bang followed by a *Turing cascade*!

10. Conclusions and Discussions

We highlighted the logical paradoxes encountered in defining FW and understanding its role in physics. We highlight our main conclusions. A principal conclusion is that like the concept of truth in Tarski's undefinability theorem, free will is also meta-theoretic: it correlates freedom to deviate from physical law in a base theory to parameters of 'will' in a meta-theory. Next, we pointed out that this meta-theoretic coupling must be continued infinitely in order to construct the FW that we intuitively associate with sentient agents. Third, the meta-theories are shown to correspond to a hierarchy of uncomputability. A consequence is that free will is uncomputable to infinite degree, in the sense that for a meta-theory at any finite order, there will deviations from its laws that are uncomputable in it. We justify this link between causality and computability by revisiting self-referential logical paradoxes that serve as the basis for theorems of Gödel-Tarski kind. Finally, we present a refinement of an anti-mechanist argument for consciousness based on these theorems.

References

1. L. Zagzebski, in *Stanford Encyclopedia of Philosophy*, ed. E. N. Zalta (2011), `plato.stanford.edu/entries/free-will-foreknowledge/`.
2. T. O'Connor(2010), `plato.stanford.edu/entries/freewill/`.
3. H. Hanaan and R. Srikanth, *On the neurological basis of free will: Locating the Gödel oracle* (2013), (under preparation).
4. C. M. Nayakar and R. Srikanth, in *Interdisciplinary Perspectives On Consciousness and the Self*, eds. A. S. Sangeetha Menon and B. V. Sreekantan, Springer (2012).
5. V. Vedral, *New Scientist* (18 November, 2006 issue), pg 55 (2006).
6. A. Zeilinger, interview in Die Weltwoche, Ausgabe 48/05 (2006).
7. G. 't Hooft, quant-ph/0701097 (2007).
8. N. Gisin, arXiv:1011.3440 (2010).

9. H. Everett, *Rev. Mod. Phys.* **29**, 454 (1957).
10. J. Conway and S. Kochen, *Found. Phys.*, **36**, 1441 (2006); arXiv:quant-ph/0604079.
11. J. Conway and S. Kochen, *Amer. Math. Soc.*, **56**, 2 (2008).
12. N. Gisin, arXiv:1002.1392 (2010).
13. R. Tumulka, *Found. Phys.*, **37**, 186197 (2007).
14. S. Goldstein, D. V. Tausk, R. Tumulka and N. Zanghi, arXiv:0905.4641 (2007).
15. A. Suarez, arxiv:1002.2697 (2010).
16. M. J. W. Hall, *Phys. Rev. Lett.* **105**, 250404 (2010).
17. A. Suarez, arxiv:0804.0871 (2008).
18. J. F. Clauser, M. A. Horne, A. Shimony and R. A. Holt, *Phys. Rev. Lett.* **23**, 880 (1969).
19. A. Suarez, arxiv:1006.2485 (2010).
20. H. B. Enderton, Stanford University, (2008).
21. M. Davis, *Computability and Unsolvability*, Dover (1982).
22. R. A. Shore and T. A. Slaman, *Math. Res. Lett.*, **6**, 711722 (1999).
23. A. Valentini, *Pramana-J. Phys.*, **59**, 269 (2002).
24. K. Gödel, *Monatshefte für Mathematik und Physik*, **38** (1931).
25. A. Tarski, *Philosophy and Phenomenological Research*, **4** (1944).
26. J. Megill, in *Internet Encyclopedia of Philosophy*, eds. J. Fieser and B. Dowden (2012), www.iep.utm.edu/lp-argue.
27. J. R. Lucas, *The Freedom of the Will*, Oxford University Press (1970).
28. R. Penrose, *The Emperor's New Mind*, Oxford University Press (1989).
29. R. Penrose, *Shadows of the Mind*, Oxford University Press(1994).
30. H. Putnam, in *Dimensions of Mind: A Symposium*, ed. S. Hook, London: Collier-Macmillan (1960).
31. P. Benacerraf, *Monist* **51**, 9 (1967).
32. C. Whiteley, *Philosophy*, **37**, 61 (1962).
33. D. M. Mackay, *Mind* **69**, 31 (1960).
34. R. Srikanth, *Uncomputability, causality and cognition* (2013), (under preparation).

Chapter 17

Mathematics and Cognition

Rajesh Kasturirangan

National Institute of Advanced Studies, Indian Institute of Science Campus, Bangalore 560 012

rkasturi@gmail.com

Mathematics is a human pursuit. Whether the truths of mathematics lie outside the human mind or emerge out of it, the actual practice of mathematics is conducted by human beings. In other words, human mathematics is the only kind of mathematics that we can pursue and human mathematics has to be built on top of cognitive capacities that are possessed by all human beings. Another way of stating the same claim is that mathematics is experienced by human beings using their cognitive capacities. This paper argues that exploring the experience of mathematics is a useful way to make headway on the foundations of mathematics. Focusing on the experience of mathematics is an empirical approach to the study of mathematics that sidesteps some of the thorniest debates from an earlier era about Platonism and Formalism in the foundations of mathematics.

1. Introduction

To a Martian, mathematics must appear to be a strange activity. What good can come out of people staring at walls and scribbling marks on pieces of paper? Some people tolerate this strange activity because it has its uses[1] — it helps us build spacecraft and figure out the riddles of subatomic phenomena. More prosaically, it helps predict the stock market. The true aficionado doesn't care about these applications; for him, the charm of mathematics lies in its purity, in the world that helps us create — a world where certainty is absolute, where truth is eternal and where deep patterns are discerned beyond the realm of ordinary humankind. Still, ordinary or not, mathematics is done by humans and so it must have some connection with our flesh and blood. Mathematics might be a lotus, but it grows out of the morass of everyday human cognition. How does that happen? What is the mental origin of mathematics, if any?

Our exploration of this question will take us through some of the thorniest questions about the nature of mathematics. This paper is hardly enough to answer those questions, but I hope to raise some interesting points about the character of mathematics — to create a beginning if not the end. It's a cognitive exploration, in that I will bring a cognitive science perspective to questions traditionally tackled by logicians, philosophers and mathematicians. I believe that mathematical thinking needs more attention from cognitive science and vice versa and hope that this paper begins to create some interest amongst cognitive scientists about the nature of mathematics and vice versa.

Let us start with the experience of mathematics. Why does mathematics appear magical to some of us? I can think of three reasons:

(1) It is purely abstract, yet completely certain. In other words, mathematical entities appear to have solidity, even though they neither have weight nor do they have density. This ability of mathematical entities to hit us over the head without having mass is one of their magical properties.

(2) It is strangely predictive of the external world outside the realm of the senses. We do quite well without mathematics when it comes to rocks and trees, but without mathematics we would be lost in the world of electrons and galaxies. Mathematics is scale invariant in a manner that everyday thinking isn't. Eugene Wigner's famous article on the unreasonable effectiveness of mathematics is perhaps the best known evocation of this aspect of the magic of mathematics.

(3) Its interconnectivity: From $e^{2\pi i} = 1$ to the Langlands program, mathematics shows us deep connections between concepts that seem utterly different.

On the surface, everyday cognition seems to lack all three properties: it isn't abstract enough, it isn't predictive enough and it isn't interconnected enough. My first goal is to show that our ordinary mind has its share of all three magical abilities too. A mathematician might say: really? What is so abstract about cats and dogs (or the word CAT and the word DOG, for that matter)? And what's so surprising about cats having four legs and politicians being habitual liars?

2. Everyday Magic

When examined closely, the everyday world is as abstract as the world of mathematics. When you look at a table from the front, you expect that it has a back as well. The table doesn't look like a half object. But no stimulation is reaching your eyes — so how do you "see" the back of the table? Everyday concepts are also abstract in other ways. For example, we intuitively feel that there's a tenuous relation between the concept CUP and cups in the world. The concept CUP has neither shape nor size nor mass, while real world cups do. Similarly, concepts interact with each other logically — we can say "can you give me either the red or the blue cup?" while objects only interact with each other causally. Cups break when they fall on a hard floor while the concept CUP does not break down when you say FLOOR.[2]

Similarly, you can talk about the past; I can say "I went to a really good lecture yesterday" and everyone understands me perfectly. But the past cannot be touched or seen. In that sense, it is as abstract as any mathematical entity. We also feel certain about the past. I know I was in that wonderful lecture yesterday with a certainty that's no less than my belief in $2 + 2 = 4$. While my certainty about yesterday's lecture might be false — my memory could be false — we do expect that most of our memories are true. We simply couldn't function in the world if our normal inferences weren't trustworthy. Consider the statement "Socrates died in 399 B.C.E." Having once existed and died, Socrates is long gone but the statement regarding his death will now be true, independent of the rest of the history of the universe. Even if human beings become extinct as a species, Socrates would still have died in 399 B.C.E and the statement regarding his death would still be true. To the extent that the human mind is a container for these kinds of entities, it is also primarily an abstract entity, whose foundational rules are abstract.

What about the existence of deep interconnections? Everyday life is full of them as well — we live in a world of regularities; otherwise the world would be unpredictable. Here are a few examples of what I would call regularities[3]:

- The size of an animal predicts the pitch of its voice. Mice squeak and lions roar and not vice versa
- Clouds are puffy while water is runny.

A particular type of regularity that's important for our purposes is the notion of an affordance.[4] An affordance is an aspect of an object that makes it "afford" certain actions from us. The lip of a cup makes it drinkable. The handle of the cup makes it graspable. Later in this paper, I will argue that numbers and other mathematical objects can be treated in the same way, that they afford features that make them useful for us.

As biological creatures, the world is transparent because regularities important for our survival have been made explicit in our minds and brains. For a deer in the savannah surprise = death. However, as anyone who has tried to design a computer system to solve seemingly obvious problems will tell you, the transparency of the everyday world hides an enormous amount of structure. The sensory world is full of interconnected structure, much of which isn't directly available to us from the world, but rather added by our mental apparatus. We make sense of the world as much we sense it.

3. Sense Making

In ordinary language, we use the word "sense" in two different ways: sense as in sensation and also sense as in making sense. The first use of sense is about perception, the second use of sense is about conceptualization. I believe that these senses of "sense" are closely interrelated; in fact they arise from principles that cut across cognition and perception. One example of a common principle is figure-ground organization. The Gestalt psychologists[5,6] thought about the figure-ground dichotomy as an organizing principle of perception. Roughly speaking, the figure is what you perceive and the ground is the background against which you perceive, though the two entities can switch back and forth on occasion (you can see images here). Cognitive linguists such as Talmy[7] have argued that the figure-ground organization works in language as well, so that spatial relations are usually conveyed with a figural object and a ground object. Consider the pair of sentences

- The red car is parked in front of the blue house.
- The blue house is situated behind the red car.

While the two sentences convey the same spatial relation, the second seems awkward, even though both of them are grammatically correct. Talmy argues that it is because figure objects are usually smaller and mobile and ground objects are larger and static. Sentence 2 flouts that rule. Talmy goes on to argue that these sentences show that spatial language also has figure-ground organization. In other words, we can assume that figure-ground is a principle of mental organization as such. It's these laws of mental organization that we need to turn to for understanding the cognitive foundations of mathematics.

4. Lakoff and Nunez

Most mathematicians are tacit platonists, i.e., they believe that mathematical entities have an objective existence outside the human world. A cognitive approach to mathematics turns that intuition on its head — it says that mathematics is what it is precisely because our minds are the way they are. In other words, the structure of mathematics is ultimately grounded in mental organization. The best example of this approach to mathematics is the book by George Lakoff and Rafael Núñez,[8] "Where Mathematics Comes From: How the Embodied Mind Brings Mathematics into Being." I will call this book LN from now on.

There are two approaches to the foundations of mathematics: internalist, i.e., foundations that approach mathematics from within subject (ála Frege and Russell), or outside, á la LN. The best known internalist example is the set-theoretic and logical foundational work of the early 20th Century,[9] with its immense and enduring impact on mathematics and other disciplines such as psychology, artificial intelligence and computer science. LN is an externalist book; they base mathematics on mental structures shared with ordinary reasoning, fundamentally based in turn on metaphorical projections of representations of space, time and objects, grounded in the activity of the brain. They reverse the earlier logicist program: logic is not the foundation of the mind; the embodied mind is the basis for logic. LN's effort is timely; mathematics is a human activity, in its foundations as well as its practice. Mathematics, like other aspects of human knowledge, needs to be grounded in human experience. However, LN answer the question "where is mathematics synthesized?" by "in the human mind", a mind unmodified by mathematics. I cannot wholly agree. Just as there is more to the world of mathematics than sets and symbols, there is more to mathematics than neurons and common sense. A mind doing mathematics meets constraints which are not simply those of the mind, though they may well be mathematical consequences of the mind's nature. While LN firmly believe that mathematical concepts are concrete, embodied and hence ultimately synthetic, their reduction of mathematics to the mind echoes efforts in the cognitive sciences that emphasize analytic knowledge.[10] Can LN have it both ways? They both extol the embodied character of mathematics, and on the other, reduce mathematics to the mind using maneuvers fundamentally tied to the mind as a repository of analytic (and thus disembodied) truths, or at least constrained by them. The phenomenological autonomy of these constraints, unaddressed by LN, is to me the core of what must be explained by any theory of "where mathematics comes from."

5. The Experience of Mathematics

Unlike LN, I start from the experience of mathematics by mathematicians. Despite an approach owing much to the phenomenological school of philosophy, LN do not address the phenomenology of mathematics itself, assuming that the mathematician's experience of doing mathematics is not important. This follows linguists (for whom all humans know grammar, but few are placed to theorize about it), but linguists' data consists of grammaticality judgments by native speakers of the language in question. What are the counterparts of these judgments when it comes to mathematics? Are there aspects of seeing mathematics from the inside that cognitivizers should be looking at?

In fact, it is the experience of mathematics that suggests a special status for its objects. From the mathematician's "first-person" point of view, the mathematical world does seem to have special properties. For example, emotional response has no obvious rôle in the evaluation or creation of mathematical concepts, while it is crucial to our use of ordinary ones. At this stage of inquiry into the synthetic foundations of mathematics, it seems important not to move too fast from the phenomenal facts to any metaphysical conclusions about the nature of mathematics. The attempt to formulate cognitive foundations of mathematics should expand rather than contract the mathematical experience.

LN's claim is that mathematics arises from everyday knowledge via metaphorical projection. A natural puzzle arises here: how we recognize a metaphor as a valid mapping. In a perceptual analogy, "what makes a metaphor veridical?" Any analysis of arithmetic as echoing the nature of object collections ignores that when doing things with numbers, we neglect most attributes (shape, color, size etc) of objects. Is there is a 'forgetful functor'[11] from object collections to mathematical sets, or a multivalued mapping whereby "3" strikes every triple? Either might embody LN's theory, but what process of abstraction it even allows us to recognize an analogy? At best the "sets = collections of objects" is a phenomenological fact, that needs to be explained by a theory of cognitive representation of sets, and cognitive representation of collections.

In the example of vision, we see three-dimensional objects. The input to our visual system includes patterns of light on the retina plus proprioception (of eyeball direction, inner ear motion sensing, etc). To say that there is a projection from the input to the output says little: the well known problem of vision is to account for the process that transforms one to the other. The transformation is highly underdetermined, and the brain's representation scheme (by which the visual cortex specifies to the motor cortex the location and shape of the toy to be grabbed) are not known. Even Gibson style 'direct perception" accounts must spell out the affordances[12] in the directly perceived world. The problems of abstraction are much harder and even more underdetermined. To state the initial and final states of the transformation (if we could describe the final states) would be to state the problem, not its solution.

These examples illustrate that a correspondence with the Category Inferences domain becomes precise only when we apply 'common sense' to the Container domain: that is, when we limit our thinking about containers to cases where the reasoning works — and as said earlier, part of the real learning process is in learning what to restrict to and when. To a mathematician this suggests, again, a mapping from category thinking (in the technical mathematical sense of Maclane[13]) rather than to it from containers, with much unreached in the container domain, though the suggestion is not a theory. The 'container' thinking is not the cause but rather the effect of an underlying category-theory-like transformation. This description reverses what LN are saying. Instead of abstract concepts being founded on concrete ones like containers, both depend on a capacity for category manipulation. That manipulation of categories is at the heart of mathematical experience.

6. Mathematics for Mathematicians Alone?

If we agree that mathematical experience is at the heart of a cognitive intervention in mathematics, then the rôle of theory is to explain its structure without presuppositions about where it comes from. What kind of theory is most compatible with it? Who are the potential theorists? A potentially embarrassment for a cognitivizer is the presence of a large group of people who care about the lived experience of mathematics, namely mathematicians. Surely, mathematicians are the most affected by their own mathematical experience and are often the best theorists of that experience. Why not leave mathematics to mathematicians?

A cognitivist response could compare mathematics with natural language. All native English speakers know English. Few reflect on its grammar or semantics until translating or teaching or formal study reveals complexities previously concealed by unconscious skill. Ordinary speakers who observe that some utterance is grammatically wrong are hard pressed to explain why. Cognitive scientists might claim (as LN indeed do) that mathematicians know mathematics but not in a theoretically insightful manner, with no more insight into the workings of the mathematical mind than the average speaker of a language knows about Broca's area. We find this deflationary account of mathematics hard to believe. Mathematicians who observe an error can be tediously explicit about its nature. Moreover, unlike language speakers who rarely reflect on linguistic processes, mathematical experience plays a major rôle in the daily life of mathematicians. Much informal mathematical conversation is about the experience of doing mathematics: what works, what doesn't, new ways of thinking about old problems, tips and perhaps the most slippery one of all — what it feels like to do mathematics with fellow mathematicians. Mathematicians regularly address all the issues of mathematical phenomenology raised earlier in this article.

Mathematicians' efforts to unify mathematics use their lived experience of doing mathematics. In two prominent 20th century examples, the unification of algebra and topology by Poincaré, Brouwer and their successors,[14] and of geometry and number theory by Weil, Serre, Grothendieck[15] and others were based on direct experience. Similarities were felt, long before the formal unifications which emerged from and validated the sense of them. One can measure theoretical insight by its explanatory power, or by its capacity to unify diverse phenomena. By either measure, mathematicians have been more successful theorists of mathematics than have psychologists. Unlike speakers of a language, mathematicians both "speak" and "reflect" on mathematics.

What can cognitive science contribute to a theory of mathematical experience? Perhaps the first role for the cognitivizer is as a clean-up act; to codify the informal aspects of mathematics not reflected in the formal structures of theorem and proof. Much informal mathematics consists of heuristics and inductive rules that later become theorems. Mathematicians (and at least one philosopher) such as Poincaré,[16] Hadamard,[17] Pólya and more recently Lakatos have all based their philosophy of mathematics on the inductive aspects of mathematical thought. Cognitive science can clearly help classify these inductive strategies, and provide experimental support for their mental basis. Cognitive scientists will be the latest avatar of an honorable tradition of making the process of doing mathematics more transparent to mathematicians and others.

Is codification of informal experience the entire role of cognitive science in mathematics, or is there something more that cognitive studies can contribute to the experiential foundations of mathematics? I share LN's vision that cognitive science can make a deep impact on the practice of mathematics, one far greater than codifying its informal rules. The hope is that cognitive science can be to mathematics what DNA and molecular biology have become to biology as a whole. However, I am not convinced that metaphor — as today understood — and other forms of ordinary conception are the DNA of mathematics. An apt analogy might be with genetics before that revolution. From Mendel onwards, scientists were aware of the laws of heredity. In the early twentieth century, scientists like M F K Fisher,[18] Sewall Wright[19] and J B S Haldane[20] formulated good genetic theories based on abstract genes and inheritance, without molecular mechanisms.

Perhaps cognitive science will first serve mathematics as Mendelian genetics served biology. The efforts of cognitive scientists and their mathematical colleagues should concentrate on bringing abstract theories to bear on phenomena like "push-back" and "right view" that are observed over the entire mathematical cosmos. The next section outlines my ideas about these abstract theories, but first I look at the impact on the cognitive sciences of applying them to mathematics.

7. Higher Cognition

Mathematics offers a rare chance to extend cognitive investigation from "normal" individuals to a revealing subset of the human population. Studies of long term meditators have changed our ideas of human limits for attention and emotional response.[21] To study mathematics and mathematicians may reshape our understanding of the limits of subitization and conceptual change. More importantly, studies of meditators and mathematicians may precipitate a methodological shift in the cognitive sciences; from studies of average individuals in normal settings to *extraordinary* individuals in *constrained* settings. Given the rich cognitive structure of mathematics, mathematical experience cannot be tucked away into a cognitive niche of 'expert behavior' like high performance athletics. Mathematical experience is a mirror for our minds, and a major function of our minds:

systematic reasoning. Precisely to the extent that mathematics aims to be about itself only, as finance and chemistry do not, the way and extent to which it is about the embodied mind — which I agree with LN is a vital truth — should show up in sharper relief. Being more polished than our normal cognition, it may offer data on the nature of the mind that cannot be obtained in the wild. Physicists can use particle accelerators to probe the extremes of the material universe. Should not cognitive scientists use mathematics to probe the mental universe?

Mathematical experience is a richly catalogued domain of human awareness related to and yet different from perceptual experience. Its qualia are unlike those of perception: the subjective experience of engaging with numbers or polygons seems different from that of seeing a red rose, perhaps more like considering its structure. Its isolation in the mind, unlike the link of perceptual consciousness to external percept, offers a possibility of well defined theories of mathematical experience. The study of mathematical experience may bring clues on the nature of consciousness as a whole. The marriage of mathematics and cognitive science offers tantalizing possibilities for a future science of the mind. With that hope in mind, I turn to certain abstract conceptions where they meet.

8. Where Mathematics Really Comes From

One of the great achievements of modern cognitive science (linguistics in particular) is the marriage of a structural question "What is the structure of knowledge in domain X?" with a metaphysical question "What are the origins of the knowledge in domain X?" From Chomsky[22] onwards, cognitive scientists have reasoned about the learnability of structure to argue for innate knowledge. The elevation of the mind as a "real" entity in modern cognitive science is deeply linked to its formulation in computational and structural terms. A computational theory of the mind makes it possible to claim that the mind is the locus where structural and metaphysical questions interact. In mentalist cognitive science, the computational mind is what we need in order to explain complex structures of behavior (as opposed to behaviorist accounts, which postulate no such entity). Mathematics certainly has its share of complex structures, which makes the general cognitive science strategy of locating complexity in an innate structure a plausible line of argument. However, I have argued above that the special properties of mathematics make the "ordinary mind" an unlikely candidate for the origins of mathematical knowledge. With this ruled out, there are two possibilities:

(a) Mathematical knowledge has a source other than the mind: a world of ideals, formal logic, or (like money) a quasi-autonomous social convention.
(b) Mathematical knowledge is neither in-house nor outsourced.

Approach (a), a natural generalization of the normal explanatory strategy in cognitive science, links the question "what is the structure of mathematics?" with the metaphysical question "where does mathematics come from?" Approach (b) rejects the "where?" question. I have argued against the "ordinary mind" as the source of mathematical knowledge. This leaves us with two sub-options:

(1) We care only about the structural question, and restrict attempts at unification to finding mathematical structures that generalize previously existing structures. This is the standard procedure in modern mathematics. Category theory has become the default formal framework in which to encode structural generalization in the practice of mathematics.
(2) We look for an alternate explanatory strategy for marrying the structural and the metaphysical questions.

Option (1), that I have called "mathematics for mathematicians alone", rejects any metaphysical commitment and focuses instead on structure. A major promise of cognitive science to mathematics is a restored focus on the metaphysical question. Taking our cognitive science lessons seriously, I seek an alternative to locus-based answers for the "structure+metaphysics" problem.

Lets start with three ideas partially inspired by the theory of affordances in Gibson[23] and by the ideas of D'Arcy Thompson[24] about the nature of form:

(i) The world of mathematics is a "real world" with objective affordances.
(ii) This world is populated by "empirical forms": "substantial" objects that occupy (a suitably generalized notion of) space, time and mind, and also have real properties and powers.
(iii) The structure of these mathematical objects is constituted by actions of many kinds, with proofs being only one these kinds.

I agree with LN that space and time lie at the foundations of mathematics, but conceptualize them not as entities internal to the human mind, but as objective aspects of the world of mathematics. In our view, the mathematical world, while special to humans, is not simply a feature of our heads. Mathematical knowledge is the knowledge of *how to act appropriately in the mathematical world.*

Like the classical set theoretic foundation builders of mathematics, LN model knowing mathematics as "knowing that" rather than "knowing how". Mathematics and the other theoretical sciences are perhaps the best candidates for "knowing that" kind of knowledge. Their truths seem abstract, theoretical, un-referred to actual experience. What if knowing a mathematical concept was much more like seeing a percept than is normally imagined? The enactive approach to perception[25] critiques the traditional model of perception based on constructing a detailed three dimensional model of the object in our minds, so that we know how the object looks like from all directions. According to the enactive approach, we don't try to build detailed representations at all. For the most part, we don't need those details, so why bother? Instead we rely on the fact that all the details of the object are available in principle, by moving our bodies around the object and pointing our eyes to the spots where we need to see more detail. That cup in the other room, I don't need to know from here what it looks like. All I need to know is that if I walk next door, I will see a cup sitting on the table. If I need to drink from it, my eyes [visual system] will guide [the motor system controlling] my fingers to it. Why should I have to represent all this detail, when I am not thirsty? Representation is expensive, and the world is always there to help me out when I need it. That is enough to know.

Most mathematicians would agree that their experience of mathematics is quite similar to this perceptual situation. Given time, most mathematicians could work through the details of Appel and Haken's proof of the 4-Color Theorem, but how many do? Mathematicians get glimpses of objects (or truths) and then — if motivated — do the hard work to "excavate" these truths from their underground matrix. If an object is particularly large or complex, it takes the joint efforts of several mathematicians, sometimes spread over centuries, to fully uncover the archeology of the mathematical domain in question. Various tools and techniques may need to be invented before excavation can commence. Over shorter time, it is particularly helpful to listen to mathematical conversation. Mathematician A may say "we can show X by compactness," mathematician B nods, and they go on to discuss the implications of X. The details of the compactness argument are no more present in their minds than the color of the teacup. However, if B says "Wait — can we really use compactness here?" they pause, settle the question (often by several stages of increasing technical detail) and return to X at a looser, more descriptive level.

The meaning of a mathematical concept resembles seeing an object, which we always encounter from a particular view. When you encounter a physical object while moving, it presents itself with a certain face, its front from where you are standing. The back is obscured, but you know it is there – you even have an inkling as to what the back looks like. Some views give a better vantage point than others, just as one can claim a "right point of view" of a particular mathematical structure. Other views may be as accurate, as technically correct, but this is a question distinct from their value in understanding. Maintaining the analogy with perception, but switching sense: surface nerves tell us about what our hands, feet and hips are touching, but proprioception tells us where they are, as parts of a whole that can work together. This does not evoke back and front masking, but it is no more a computer-style neutral 3D model with all corners listed than is the visual percept. The brain maps some parts of the body in immense detail, others sketchily, and pragmatically varies attention over the body and time. Mathematicians understood that homology theories resemble each other before Eilenberg and Steenrod axiomatized them. After that they knew that anything that testably "is a homology theory" will give corresponding results on major classes of topological space, like polyhedral meshes and cell complexes. This feels like a correct, central perception to replace the particular constructions and lemmas by which the individual systems were known.

These perceptual analogies also explain why mathematical knowledge seems more like discovery of empirical facts than a statement of tautological truths. We don't know the back of concepts (or objects) in the way we know their fronts, or our faces in a mirror. An object's back is an absence that is vividly present to our perceptual apparatus, just as for the meaning of a mathematical concept. It has hidden layers of meaning and detail, but we know that they exist, in the more-than-existential sense that they are enactively available to us. We hate to commit ourselves to what a mathematical object or concepts looks like in the round just by studying the front: "I can see a sheep, at least half of which is black." The meaning of a mathematical concept stays open to other possibilities depending on your view of it, even if you have a strong bias to thinking about it in one way. The open endedness of mathematics is a natural consequence of the fact that the meaning of a mathematical concept is not entirely available in one perception.

Can one make this sense of a mathematical concept as "object in a mathematical world" into a precise theory? This runs into deep conceptual problems related to the nature of mathematical objects (if indeed they are true objects). Gibson could say that the objects of perception are "out there in the world". It is not clear what kind of world contains mathematical objects and what might be the actions that constitute them. I do not offer an answer to that question, but I do not find it more of an obstacle than a 17th Century physicist's limited knowledge of the nature and constitution of the physical world. Great progress in the physical sciences assumed a physical world "out there", independent of human concerns. Similarly, we might have to postulate a human world out there, independent of my individual concerns but constrained by my biology that serves as the ultimate ground of mathematics. That will leave us with a modified version of Wigner's question:

How come the human world is so good at accessing aspects of the non-human world?

That's a question well beyond the scope of this paper, but with any luck, it will give us a useful way of restating the magic of mathematics while embedding that question within a larger world of human concern.

9. Conclusion: From Greek to Indic Philosophies of Mathematics

Let me end this article with a speculation about philosophical approaches to mathematical experience. One reason why LN's approach to the cognitive foundations of mathematics seems déjá vu all over again is its place in a tradition where embodiment and experience are less important than innate, analytic structures as sources of knowledge. In this tradition, running from Plato to Chomsky, mathematics is an "ideal" form of knowledge, with standards of rigor and certainty only be dreamt of by other disciplines. Despite recognizing the dangers of idealizing mathematics, LN share the same metaphysical biases.

If one is serious about moving the foundations of mathematics away from logic and abstraction to more experiential forms of knowledge, it might be better to start with a philosophical tradition that takes experience more seriously than the dominant tradition of western philosophy. More broadly, we would like to give the mathematician's activity a more central rôle, taking an Indic approach to mathematics. The Greeks concentrated on static truths, such as the Pythagoras Theorem, while their Indian contemporaries took procedures as central.[26] There was of course overlap, as illustrated by the Euclidean Algorithm and geometric construction procedures, but with a major difference of emphasis. LN take an essentially Greek approach, in telling us what a metaphor 'is': not only would we like more detail on this, but we would prefer to know how an embodied metaphor acts, what procedure is followed by the brain — or at least to have experimental data that powerfully constrain our theorizing about what procedure may be embodied in brain activity. The static characterizations that we get from Euclid or from today's logicians may obstruct our grasp of the brain's relations to mathematics.

Such subfields of cognitive science as the embodied and enactive have already somewhat challenged these static assumptions. LN's attempt to break from classical theorizing about mathematics uses embodiment as a tool to chip away at some of the older presuppositions. However, situating their challenges within a matrix of logicist thinking makes the pull of logicism much stronger in LN's cognitive foundations of mathematics than conceptually necessary. This holds especially when seen from the point of view of Indian traditions that accept embodiment as a given. Even Indian logic is empirical: the basic schemas for correct reasoning incorporate empirical elements, and a syllogism requires an example. Even a "logical foundations of mathematics" with roots in Indian logic will be more synthetic, more experiential and more procedural than a "cognitive foundations" derived from euro-logicism. The cognitive foundations of mathematics will become truly cognitive and truly phenomenological only when the mathematical experiences of multiple cultures are taken into account.

References

1. Pickover, Clifford. *The math book: from Pythagoras to the 57th dimension, 250 milestones in the history of mathematics.* Sterling Publishing, 2009.
2. Macnamara, John. *Through the rearview mirror: Historical reflections on psychology.* The MIT Press, 1999.
3. Alexander, R McNeill. *Optima for animals.* Princeton University Press, 1996.
4. Gibson, James Jerome. *The ecological approach to visual perception.* Routledge, 1986.
5. Köhler, Wolfgang. *Gestalt psychology: An introduction to new concepts in modern psychology.* WW Norton & Company, 1970.
6. Koffka, Kurt. *"Principles of Gestalt psychology."* 53 (1935).
7. Talmy, Leonard. *Toward a cognitive semantics, Vol. 1: Concept structuring systems.* The MIT Press, 2000.

8. Lakoff, George, and Rafael E N Núñez. *Where mathematics comes from: How the embodied mind brings mathematics into being.* Basic books, 2000.

9. Van Heijenoort, Jean. *From Freege to Gödel: A Source Book in Mathematical Logic,* 1879–1931. Jean Van Heijenoort. Harvard University Press, 1977.

10. Chomsky, Noam. *Knowledge of language: Its nature, origins, and use.* Greenwood Publishing Group, 1986.

11. Mac Lane, Saunders. "Category theory for the working mathematician." *Graduate Texts in Mathematics* 5 (1971).

12. Gibson, James Jerome. *The ecological approach to visual perception.* Routledge, 1986.

13. Mac Lane, Saunders. "Category theory for the working mathematician." *Graduate Texts in Mathematics* 5 (1971).

14. Dieudonné, Jean. *A history of algebraic and differential topology,* 1900–1960. Springer, 2009.

15. Grothendieck, A, and J Dieudonné. "Éléments de géométrie algébrique." *Publ. math. IHÉS* 8.24 (1965): 2–7.

16. Poincaré, Henri. *Science and hypothesis.* Science Press, 1905.

17. Hadamard, J, "An Essay on the Psychology of Invention in the Mathematical Field", Dover, New York (1954).

18. Fisher, Ronald Aylmer. *The genetical theory of natural selection: a complete variorum edition.* Oxford University Press, 1999.

19. Wright, Sewall. "Evolution in Mendelian populations." *Genetics* 16.2 (1931): 97.

20. Haldane, John Burdon Sanderson. *The causes of evolution.* Princeton University Press, 1990.

21. Lutz, Antoine *et al.* "Attention regulation and monitoring in meditation." *Trends in cognitive sciences* 12.4 (2008): 163-169.

22. Chomsky, Noam. *Knowledge of language: Its nature, origins, and use.* Greenwood Publishing Group, 1986.

23. Gibson, James Jerome. *The ecological approach to visual perception.* Routledge, 1986.

24. Thompson, Darcy Wentworth. "On growth and form." *On growth and form.* (1942).

25. Varela, Franscisco J, Evan T Thompson, and Eleanor Rosch. *The embodied mind: Cognitive science and human experience.* The MIT Press, 1991.

26. Raju, CK. "Computers, mathematics education, and the alternative epistemology of the calculus in the Yuktibhâsâ." *Philosophy East and West* 51.3 (2001): 325-362.

Index